OVERCOMING BIAS HABITS

OVERCOMING BIAS HABITS

An Evidence-Based Guide to Creating a Joyfully Inclusive World

WILLIAM T. L. COX, PhD

UNIVERSITY OF CALIFORNIA PRESS

University of California Press
Oakland, California

Library of Congress Cataloging-in-Publication Data

Names: Cox, William T. L., 1984– author
Title: Overcoming bias habits : an evidence-based guide to creating a joyfully inclusive world / William Taylor Laimaka Cox.
Description: Oakland, California : University of California Press, [2026] | Includes bibliographical references.
Identifiers: LCCN 2025034994 (print) | LCCN 2025034995 (ebook) | ISBN 9780520410411 cloth | ISBN 9780520410428 paperback | ISBN 9780520410435 ebook
Subjects: LCSH: Prejudices—Prevention | Stereotypes (Social psychology)—Prevention | Social change
Classification: LCC HM1091 .C69 2026 (print) | LCC HM1091 (ebook)
LC record available at https://lccn.loc.gov/2025034994
LC ebook record available at https://lccn.loc.gov/2025034995

Manufactured in the United States of America

GPSR Authorized Representative: Easy Access System Europe, Mustamäe tee 50, 10621 Tallinn, Estonia, gpsr.requests@easproject.com

35 34 33 32 31 30 29 28 27 26
10 9 8 7 6 5 4 3 2 1

Contents

Acknowledgments

THE MAJORITY OF THE STORIES herein are not mine—I'm just the one who wrote them down. The stories belong to the people who lived them and were gracious enough to share them with me, and through me, with you, the reader. So, first and foremost, I must thank the many people whose firsthand experiences enabled me to translate this book's scientific content into relatable, practical stories to help us all create meaningful changes in our lives. I also want to thank everyone who has ever participated in our training program; even if their specific stories did not make it into the final draft of this book, their insights and reactions to our training over the years have helped me hone all the content to be maximally useful and impactful.

I like to think I've had a clever idea or two in my career, but good science lives not in any one person's cleverness, but in the cumulative, painstaking work of many, *many* scientists. Nothing I've distilled in this book would be possible without decades upon decades of trial and error, theorizing, hypothesizing, and refinement by more scientists than I could hope to name or cite, across many disciplines. And the science of human behavior in particular depends on the hundreds of thousands of people who graciously agree to serve as human research participants. I want to sincerely thank my past and present scientific colleagues and the many participants, in both my own research studies and those of other scientists, who make the research possible.

No one shapes a scientist's career more than their PhD advisor, and I had one of the greats! I'd like to thank my advisor, Dr. Trish Devine, not only for guiding and molding me throughout graduate school, but for the many great years of work we continued after I graduated, as we developed much of what became this book's content, both in the lab and on the road! I also want to thank the many other colleagues with whom I developed, tested, and discussed our training program over the years. I'd also like to thank a few specific colleagues, whose influence powerfully shaped me and my thinking in ways that contributed to this book: Dr. Janet Hyde, who helped hone my scientific perspective on tackling social justice issues; Dr. Lyn Abramson, with whom I developed the theoretical framework to bring in evidence-based clinical therapies as methods to reduce bias; and Dr. Amber Nelson, who helped me apply and adapt those clinical therapies, and, importantly, who brought into focus the central role of infusing *joy* into diversity work.

I'd like to thank my parents, family, chosen family, and friends for their companionship and support in life generally, and as I wrote this book specifically, with a special shout-out to Christine Heilman for publishing advice and help editing. I'd especially like to thank my husband, Eric Roman Beining, who was the constant victim of my brainstorm bombing and was this book's first reader and most vicious editor. I'd like to thank my actual editor, Chloe Layman, who first sought me out to write this book, and then skillfully and patiently shepherded me through every phase of publication. I'd also like to thank my insightful peer reviewers and the many people at University of California Press, who helped at every phase of publication to make this beautiful final product.

Lastly, I want to thank myself, for the many years of learning, tough choices, determination, and perseverance as I worked to harness science in service of effecting practical change in the world, which has now resulted in this book.

UNIT ONE

THE SCIENCE AND PRACTICE OF MEANINGFUL CHANGE

EVERYTHING IN THIS BOOK is a journey geared toward empowering you to be an impactful agent of change in the realm of bias and diversity, and the chapters in this unit provide the road map for that journey! All the content in this book is based in well-replicated science, but that science has been translated into very **practical, action-oriented skills** that you can start using to create meaningful change in your life. We're going to learn how race bias, gender bias, and other forms of bias are types of mental habits, and our approach to reducing bias will involve learning how to break those bias habits. To counteract the anxiety and fear some folks feel around diversity topics, we're also going to focus on infusing *joy* into our change efforts. This unit will explore these entwined goals of learning to **disrupt bias habits** and **cultivate diverse joy** as we start our journey.

INTRODUCING OUR JOURNEY

Bias Habits and Diverse Joy

HAVE YOU EVER slipped up and made an unfair assumption about someone because of their race, gender, or other group status? Have you ever felt unsure what to do when someone says you or your organization needs to "do better" with diversity, equity, and inclusion (DEI)? Have you ever attended a diversity training or bias training that left you feeling powerless, defeated, or ill-equipped to make progress? If so, you're not alone! *Most* people recognize that intergroup biases, inequities, and disparities are major problems, and they often turn to diversity training, hoping to find real solutions. Scientific research indicates, however, that by and large, diversity trainings fall short of consistently delivering meaningful change.

Does that mean we just give up, because there is no hope to make progress on bias and diversity problems? Well, no! Of course not. We just need to use science! The problem isn't that there *can't* be an effective approach to diversity training; the problem is that it needs to be done well, which means that trainings should be based on a solid scientific foundation and then be rigorously tested and refined in randomized controlled experiments. The goal of understanding, predicting, and changing human behavior is best served by the scientific method, and addressing issues of bias, diversity, and inclusion is no exception; it requires a scientific, evidence-based approach to create meaningful change.

Inspiring lasting, meaningful changes in human behavior is no small feat, and behavioral scientists spend their whole careers identifying how to empower change. I am one of those scientists—my life's work has been devoted to harnessing the science of cognitive behavioral change to empower people and organizations to enact real-world bias and diversity solutions. For almost two decades, my colleagues and I have worked to develop, test, and refine an evidence-based diversity intervention called the *bias habit-breaking training*. Extensive, rigorous scientific testing has shown that this program **empowers people to make positive changes** in themselves and their organizations, working hard to reduce bias, create inclusion, and promote equity. This research has also shown that people who go through our training and then put into practice what they learned **have less mental bias**, **stereotype less**, and **speak up more** about diversity issues. In organizations that receive our training, we've observed significant improvements in organizational climate and increases in the hiring of members of underrepresented groups.

My colleagues and I have spent years refining this intervention as an in-person workshop, and now I've translated it into this book. I want you to think of this book as a journey we're taking together to understand how **each and every one of us can be an effective agent of personal and social change**. You might read this book solo, in which case I will be your primary companion on this journey, or you might consider using this book as a group activity to discuss and learn with others.

A BIT ABOUT ME

In some ways, who I am should not matter—the primary reason to trust what I say in this book is because of the science behind it. But this book is designed to be a journey we take together, and I'd like you to know a little about me, your travel partner.

Issues of culture and nationality, race and gender, and stereotypes, prejudice, and bias were built into the fabric of my life. As one of five adopted children in a military family, I grew up moving both around the world and across the United States. With each new place I lived, I found that people had different habitual ways of seeing and interacting with the world around them. These distinct cultural habits were not only apparent across different countries or continents, but even across different communities within the United States, and I came to realize that if I wanted to understand someone, having knowledge of their cultural context and their habitual way of seeing the world would help me.

My family itself is extremely multicultural and multiracial: my mother is Native Hawaiian, Chinese, and Portuguese; my father and two of my brothers are White; and two of my siblings are of Puerto Rican descent. From a very young age, I observed how my family members' broad spectrum of skin color influenced how they were treated. My parents are also conservative Mormons, which brought into focus prejudice against LGBTQ+ identities. Soon after I came out to my parents as a gay man at seventeen, they threw me out of their home. I was initially homeless, but then secured a job and housing. I worked a full-time job through community college and university, and ultimately completed my PhD and have had an award-winning scientific career.

These experiences from my life not only provide examples of how bias and prejudice happen, but also showcase how we can move in positive directions. One of the most powerful of these examples is how, over the past twenty years, I have worked to rebuild a relationship with my parents and help them progress from where they were when they threw me out for being gay to today, where they accept me and welcome my husband as part of the family. These and other firsthand experiences with issues of race, gender, LGBTQ+ identities, culture, and poverty powerfully shaped me, and ultimately, the direction of my work. My life experiences gave me a passion for combating prejudice and bias, and championing inclusion and diversity.

I have always been fairly scientifically minded, and in college I was fascinated to learn there was a whole field devoted to experimentally, empirically studying bias, prejudice, and stereotyping. I built this fascination into a career, completing my PhD in social psychology and devoting my life to the scientific study of stereotypes, biases, and inequities—how they operate, why they persist, and, most importantly of all, how to help people overcome them.

Half my scientific career involves *basic* science, which is focused on understanding the fundamentals of how things work. I've published research on how culture perpetuates stereotypes, how your brain's learning mechanisms contribute to the persistence of bias habits, and how biases in your mind can lead to discriminatory behaviors. But I didn't become a scientist just to *study* problems related to bias and diversity; I wanted to help find *solutions* to those problems. How do we take what we know about bias, disparities, and other diversity issues and use it to build a better world? This question brings us to the other half of my scientific career, which is focused on *applied* science.

My many years of applied science involved developing and refining the bias habit-breaking training, which is the basis for this book! Central to this training program is that it translates science into practical skills, teaching people that biases are like "habits of mind," and the process of reducing bias involves learning to disrupt those **bias habits**. Importantly, over the past two decades, my colleagues and I have tested this training program using several large randomized controlled experiments—the same sort of testing drugs and medical devices go through—because that's the only scientifically valid way to test whether something *causes* changes in important outcomes. I'll talk more about these outcomes throughout this book, but in brief, the training program empowers people to take charge of their own minds, behavior, and environments to reduce bias and cultivate more inclusion, belonging, and equity. Of course, doing such large research studies requires an amazing team, and I've had many wonderful collaborators in this work over the years. Two in particular I'd like to mention are my PhD advisor, Dr. Trish Devine, who originated the training's central guiding theme of approaching bias as a habit to be broken, and Dr. Molly Carnes, who obtained our first large research grant to test the training.

In response to several of our research publications, major media outlets—including NPR, the BBC, *Harvard Business Review*, and *The Atlantic*—have covered our work or published interviews with myself or my colleagues about the training. Following some of this mainstream media attention, we began receiving requests from organizations around the world who wanted to receive the training. Given the documented shortcomings of nonscientific diversity trainings and the body of evidence showing our training's effectiveness, my colleagues and I decided to disseminate the bias habit-breaking training for these audiences, independently of our ever-ongoing scientific testing and refinement of the training.

Over the past decade, I've delivered the training to thousands of people in over a hundred audiences all around the US and the world. These audiences came from just about any field you can think of (e.g., police officers, state legislatures, academics, preschool teachers, lawyers, medical professionals). That extension of our work—delivering the training live, "out in the world," unconnected to any formal research study—transformed how I think about my work, and myself. It was an unexpected new role for me; I was no longer just a scientist, but a practitioner as well. For a scientist conducting an experiment, the ideal is to be disconnected from the participants as much as possible. In fact, in most of our research studies, the training was delivered using a voice recording or a video of me, so there was no chance for interaction with the audience. But when we started doing this practitioner work separate from the experiments, actually going to organizations and doing interactive work with people, it was transformative to see firsthand the impact our work had for so many people. It led me to start seeing myself as not just a scientist, but a scientist-practitioner.

I still did and do all that wonderful scientific work, including ongoing experiments testing updated versions of our training program. But the practitioner side of the work became equally important to me, really seeing how our training was touching people's lives, working with them on bias and diversity problems, and helping them apply the skills our training teaches. And that practitioner work *also* informed my scientific work, as I learned directly about the pervasive real-world problems people face, rather than just understanding them via research studies. My integration of the scientist and practitioner roles made both sides of the work stronger.

INFUSING *JOY* BACK INTO DIVERSITY EFFORTS

Through this work as a practitioner, I've witnessed how so many people in every corner of the professional world are exasperated with nonscientific diversity efforts (which is also what the scientific evidence says about people's response to run-of-the-mill diversity efforts). They were tired of DEI efforts just being words, not actions. Our training, in contrast, gives people actionable skills they can use to make real, meaningful change for themselves and their organizations. Unexpected, however, was seeing how much the training brought people hope and *joy*.

I could see how the practical skills our training taught them gave people hope and a path forward to make progress. It started making diversity something joyful again. I say "again" because I believe that for so many people, myself included, diversity started as a source of joy, but then that joy was stripped away by recent toxic, politicized rhetoric around diversity and DEI, which only seems to be escalating. Growing up as a military kid, whenever my family moved somewhere new, I was excited to learn about the different cultural habits and ways of doing things. Or when I'd meet someone with a different background from me, it was interesting to see all the ways they saw the world similarly to me, as well as the ways they saw things differently. I believe that diversity can, and should, be something joyful—and one way you could reframe our training program is to say that what it does for people is help them overcome bias habits that are obstacles to that joy.

Bias, in its many forms, **robs people of joy**. When someone messes up and unintentionally says or does something biased, it steals their joy because afterward they feel guilty or upset that they've made a mistake and perhaps caused harm to someone. And the person harmed by the bias is also robbed of their joy when they feel like they've been slighted or disrespected. When people lack practical, effective ways to address bias, they start worrying and freezing up, because they're scared of doing something wrong but feel powerless to know what they should do instead. The cultural concept of *diversity* becomes a source of fear, panic, and anger, rather than one of joy.

In recent years, I've worked with my close friend and colleague, Dr. Amber Nelson, to develop this concept of cultivating **diverse joy** as essential to bias reduction, building on the established foundation of the bias habit-breaking training. This work epitomizes the benefits of my scientist-practitioner approach. As a practitioner, I've observed the role joy has played in people's efforts to overcome bias; as a scientist, I can draw on the scientific literature to explain and enhance the role of joy in this work. We know that people learn better from approaching something pleasant (i.e., joyful) than avoiding something unpleasant. If you want people to stay motivated to work on something over the long term, the two best ways to help them maintain that motivation are

Figure 1. Our high-level goals on this journey have reciprocal benefits.

to make the work feel useful and relevant (as our training does by teaching practical, action-oriented skills) and to make it enjoyable. In fact, many researchers even *define* bias as people having more pleasant (i.e., joyful) associations with people in their own social groups compared to other groups, with some going so far as to say that this "in-group favoritism" is the *primary* force driving bias processes.

Odds are good that many of your deepest, most joyful memories aren't very diverse—you likely have more joy associated with people similar to yourself. There's nothing morally wrong with that default state—for most people, it's merely a byproduct of growing up in a family and community in which more people are similar to you! But one of its unintended consequences is bias. And a major way to work against that bias is to cultivate more diverse joy. When you have more positive, joyful interactions with people who are different from you, it puts associations in your brain that make it less likely you will fall prey to biases related to those group statuses. When you become close friends with someone different from you, the emotional connections actually work against bias habits in your mind. So, as we go on this journey, I want you to keep both of these intertwined goals in mind. We're going to

create meaningful change by developing skills to *both* disrupt bias habits *and* cultivate diverse joy.

This book is the culmination of my life's work so far. If you take it seriously, engage with its content, and put effort into applying what you learn, this book can truly empower you to **be an impactful agent of change**, just as our bias habit-breaking training has done for countless others, in my work as both a scientist and a practitioner.

SOME INSTRUCTIONS AND GUIDANCE

I want to acknowledge right off the bat that **the topics covered in this book—bias, racism, sexism, prejudice, and oppression—can be difficult to talk about** and can bring up uncomfortable emotions. You may experience discomfort as you read because you may remember times when you expressed bias and felt guilt about it. Or possibly it will make you think of how a bias could disadvantage you or people from your social group. These are valid emotions. But I want to assure you that when these topics come up and have this type of emotional baggage, it's not just for the sake of making you feel bad. The point isn't to spend time wallowing in despair, shame, blame, finger-pointing, or anything like that. Whenever I talk about a difficult topic, it's always with an eye on **moving forward to make a change for the better**. Creating change usually involves some level of discomfort, as we learn new ways to do things.

The fundamental goal of the bias habit-breaking training is to empower you to be an impactful agent of change related to race bias, gender bias, and other forms of prejudice, bias, and inequity. Research consistently shows that *most* people—nearly everyone—hold fairness and equality as very strong personal values and therefore oppose bias. Biases persist not because we want them to, but because we don't always have the skills to recognize and correct them effectively. As I will discuss throughout our journey together, biases are like bad habits that can influence our behavior unintentionally, but we can develop skills to break those bias habits! This training program focuses on helping **develop those bias habit-breaking skills**, in service of our intertwined goals to disrupt bias habits and to cultivate diverse joy.

This book is science-based, but it is not a science textbook; it will feel more like a thoughtful conversation than a science lecture. Together, we will go on a journey of self-reflection, problem solving, and skill development that is strongly based in well-replicated science, but all the science has been translated into highly practical, actionable steps that you can immediately apply in your life. You'll learn about each concept or skill first as a general principle, then I'll walk us through a handful of real-world examples showing how others have applied that skill to make meaningful changes in their lives, so you get to benefit from the work of others who came before you!

Everything on this journey is **action oriented** and **skill based**—you'll learn to recognize and identify specific ways bias habits might play out, then obtain tools you can use to disrupt those bias habits. You'll learn about addressing these issues at **the individual level** (learning to reduce bias in your own mind and behavior), **the interpersonal level** (learning to create inclusion and belonging), and **the institutional or systemic level** (learning to promote equity). You'll also be able to apply what you learn broadly, to any type of bias, whether related to race, gender, sexual orientation, disability, age, or any other social group status.

Each chapter concludes with a **Reflection Items** section designed to help you apply what you've learned and connect it to your daily life. Your own reflections and insights are a key part of the training program's process; the impact this book will have for you depends on what you put into it. When it comes to your own life and experiences, no one is a better expert than you, and only you can see when and how to apply what you learn here. The reflection prompts are designed to help you take a moment and do that work, rather than just moving directly to the new skills in the next chapter.

This book was designed so that you can experience the training either on your own or with a group, like a book club, or as a workplace activity. If you're reading this book with a group, use the reflections as well as the resources on BiasHabit.com/book as jumping-off points to discuss these topics with each other. If you are reading this book as a solo activity, I encourage you to pause and make use of these items before moving to the next topic. You're not just reading to learn a list of facts; the book is an *exercise*, meant to prompt you to ponder and reflect. Use the reflection items to consider how the material and principles covered in each chapter may play out in your life.

Whether experiencing this training on your own or with a group, you can also think about whether you want to complete it all in one sitting or one chapter at a time. Each chapter is written to get you to a point where you can do some serious reflection or discussion, so a reading group might consider, for instance, discussing one chapter a week. I strongly advise against skipping chapters, however, because each and every chapter provides part of the foundation that later chapters will build upon. Even if you already know about the core idea of a given chapter (e.g., most people have heard about the idea of a "self-fulfilling prophecy," as in chapter 3), the point of this book is not to simply relay a set of facts; it's about developing skills to put that knowledge to work in your daily life. If we stripped everything in the training down to just the simple facts without translating them into practical skills, this wouldn't be a book, it would be a short handout of terms and definitions, but you wouldn't have the skills to *apply* that knowledge.

FEATURES OF BIAS HABIT-BREAKING SKILLS

I'd like to highlight a number of key features of the skills to keep in mind as we go forward. Each of the bias habit-breaking skills is designed to be **actionable**, **self-sustainable**, **generalizable**, and **customizable**.

Actionable

Everything in this book is geared toward developing *actionable* skills. Knowledge that just sits in your mind but fails to translate into actually *doing* anything differently won't help you in your efforts to make meaningful change. As you read, ask yourself, **"What will I *do* with this information?"** Sometimes the actions will be overt and obvious, like going to new and different types of events to expose yourself to new ideas or speaking up when you see a problem. Some will be more private, like tuning in to emotional reactions, doing mental exercises, or learning to notice and evaluate messages you're absorbing from the media. But nothing herein is shared as knowledge merely for knowledge's sake—there is always some form of action with which to engage.

Self-Sustainable

The skills are all meant to be *self-sustainable* things that you as an individual can maintain over time. Nothing you'll learn will be as simple as "flipping a switch," in which one quick change solves the entire problem for good. These skills are things you will practice over time. I'm not a fan of the phrase "practice makes perfect," because I believe perfection is neither achievable nor desirable. Instead, I prefer to say that **"practice makes progress."** As you learn these skills, keep your mental eye on how you can maintain and apply the skills over time. With practice, they'll get easier and more natural, and you'll find new ways they can help you. Further, how you apply a skill may change over time, as you gain new insights or as culture or your context changes. This is not a book of easy and immutable answers, but a guidebook of skills to find, develop, and revise your own answers over time.

Generalizable

The skills are *generalizable* across various forms of bias (i.e., biases related to race, gender, LGBTQ+ status, social class, disability, age, politics, body size, and any other groups or intersections of groups you can think of). Everything we're learning taps into *general principles* of how your brain works, how interpersonal interactions unfold, how organizational practices operate, and so on. If I illustrate a principle using a gender bias example, I don't want you to think what we're learning is *just* about gender bias. I want you to **think beyond the specific examples** and **develop each skill** to see how it can apply to any and all types of bias you can think of.

Customizable

A final important feature of these skills is that they are *customizable*, meaning they can be adapted for many different contexts in your life. As these skills tap into general principles of human experience, they are just as applicable for lawyers as they are for schoolteachers, doctors, police officers, and anyone else making their way through life as a human! I'll use examples drawn from these and many other contexts to share how others have adapted and customized the skills for their lives. But it's up to you to fit these skills into your life! **No one is a better expert on your life than you!** And if you have a job, no one knows the ins and outs of your job better than you do. Even within a single profession, the daily procedures and policies and workflows can differ from one person to another or one location to another. I cannot know your own life or job as well as you do, but these skills are designed for you to be able to customize them into many different contexts.

REFLECTION ITEMS

- Have you ever been to a diversity training that left you feeling powerless or you felt was a waste of time? Or one that you found to be helpful and useful? What worked or didn't work in those trainings?
- Many nonscientific approaches to diversity work end up making people feel hopeless to make any kind of meaningful change or afraid to engage with diverse people or diversity topics. Have you ever encountered or experienced this? How might people discuss those issues in a more productive way?
- I introduced the concept of *cultivating diverse joy* as one of our high-level goals in this book. Think of a time you found joy in diversity or shared joy with someone different from yourself. What did it mean to you, and can you think of a way that joy might play a role in reducing bias?
- Make sure you understand the key features of the skills I laid out. In the reflection items at the end of future chapters, I'll prompt you to consider how that chapter's content is *actionable*, *self-sustainable*, *generalizable*, and *customizable*.

SHARE YOUR REFLECTIONS!

As you engage with this book and its reflection items, you will come up with your own examples of how bias habits operate and your own ways to apply the bias habit-breaking skills to disrupt those bias habits. Just as you will benefit from other people's examples in this book, my research team and I hope you'll help others by sharing your examples and experiences! Your own examples and experiences can contribute to our research, help us improve this training program, and provide guidance to others who complete this training in the future! If you're willing, please **share your own examples and insights** with us. If we ever use an example you shared with us, we will of course anonymize it to protect your privacy. You'll see that most of the real-world examples throughout this book are anonymized using fake names, except in a few cases where I provide the first and last name, because the example involves a public figure or someone from my life who gave me permission to name them. You can send us your examples and thoughts through a form on our website, at BiasHabit.com/share.

RESOURCES

I founded a nonprofit, Inequity Agents of Change, with the express mission to provide free evidence-based diversity resources, built on the foundation of the bias habit-breaking training. Those resources, therefore, can help you on your journey in this book! For some of these supplemental resources for each chapter, check out BiasHabit.com/book.

- *Review Videos.* For most of the key concepts in the training, there are short review videos where I **define the core skill** and **provide some additional examples** of how to use it. If you're reading and discussing this book with others, these brief videos can be useful to refresh people on what they read to help get your discussion started.
- *Further Readings and Examples.* For many of the chapters, there will be links to additional readings and resources if you'd like to **learn more** about something covered in the chapter. As readers like you share their reflections and insights with us using the BiasHabit.com/share link, we'll also provide some of those examples there.
- *Podcast Episodes.* Every chapter has at least one podcast episode to accompany it. The podcast, ***Diverse Joy***, was specifically

developed to be an extension of the bias habit-breaking training and a way to keep the learning and work going after the training session. I cohost the podcast with my friend and colleague Dr. Amber Nelson, who is a clinical psychologist. Much of the bias habit-breaking training was developed and enhanced using insights drawn from evidence-based clinical interventions (e.g., cognitive behavioral therapy); in our podcast, Amber brings in even more of those insights from her work as a clinical practitioner. As the podcast's name implies, we focus on **bringing *joy* into diversity discussions**, making the podcast both entertaining *and* educational. In each episode of *Diverse Joy*, there is a segment devoted to reviewing one of the bias habit-breaking skills and expanding on it. On the resources web page for each chapter, we'll provide links to the specific episodes that correspond to that chapter's skills. The podcast was designed to be able to stand alone, however, so you can also just listen to the podcast at your own pace, without lining up episodes with each chapter you read. Either way, find it by searching for "Diverse Joy" in your podcasting app, check it out on our website at DiverseJoy.com, or watch the video version on YouTube, where you get to see Amber and me as we talk, in addition to hearing our voices!

1

BIAS HABITS AND HOW TO BREAK THEM

EVERYONE STARTS LEARNING biases from a very young age. Kids as young as four already know predominant stereotypes and generalizations about major social groups. Kids that age don't have the word *stereotype* in their vocabulary, but they'll say things and act in ways that reveal the ideas they've learned. For instance, many parents observe their young boys trying to get out of doing chores with excuses such as "That's girl stuff!," showing they've already learned that cooking and cleaning are more associated with women than men.

In general, parents do not explicitly teach or encourage kids to learn these ideas. Kids pick up on stereotypes from culture, the media, and their social environments. They observe and learn these stereotypes about many different social groups. A parent shared with me a story about when they noticed that their young son started calling other kids "gay" as a derogative. When they asked him what he meant when he said that, he replied with a fake lisp, "Well, gay is when a guy is acting like *this*" as he pantomimed someone walking and talking in a hyperfeminized, swishy way—like an extremely flamboyant stereotype of a gay man. When the parent discussed this further with him, they found out he had no idea that *gay* referred to men who are attracted to other men. In other words, he knew the stereotype of the group before he even knew what the group label actually meant.

Another time, a father told me about an instance when he took his five-year-old son to a doctor's appointment. When the child noticed a Black family sitting in the waiting room, he pointed at them and started singing the theme song from the TV show *Cops*. The father, who was mortified by his child's behavior, realized that because his family was White and lived in a predominantly White neighborhood, *Cops* was the main, or perhaps only, time his son was exposed to Black people. His son therefore learned that when people have darker skin, they are criminals, because that's the only context in which he saw people of color.

CULTURAL INPUT TEACHES BIAS HABITS

Stereotypes about social groups become ingrained at very young ages, before kids' brains have developed the regulatory capacity or moral reasoning capacity to *question* these ideas. This means that **stereotypes become the default, habitual, automatic way of perceiving other people**—for them as kids, and then for us as adults. We continue to be bombarded with stereotypical images and messages throughout our lives, continually reinforcing these ideas in our minds.

Many aspects of our social environments teach us stereotypes, but media is one of the biggest contributors: TV shows, commercials, movies, books, social media, and more all portray people in ways that ingrain stereotypes in our minds, even when we're not giving the media our full attention. Even as modern media become more thoughtful and diverse, old media (e.g., classic movies, TV shows, books) that reflect the biases of their times remain a part of our media and cultural landscape. Once a piece of media enters the zeitgeist, it tends to persist—it has ***cultural inertia***, even when it includes portrayals we may find offensive by modern standards. We cannot shield ourselves from all the cultural input that contains stereotypes and biases, but you can start to **monitor your input** and pay attention to what messages your mind is absorbing from the media and culture.

For many years, I've shown people a stark example of race bias involving two newspaper clippings about 2005's Hurricane Katrina. Both articles were published the same day, reporting the same type of event, with very similar photographs, and captioned using nearly identical sentences. The first shows a young Black boy who appears to be around twelve years old, carrying some supplies, with the caption, "A young man walks through chest deep flood water after looting a grocery store." The other article's picture shows two White people, and its caption reads, "Two residents wade through chest-deep water after finding bread and soda from a local grocery store." The Black boy was characterized as "looting," but the White folks were "finding" their supplies. The caption doesn't say the White folks *bought* their supplies—if you "find" something without paying for it, that is also looting. But their behavior wasn't characterized as such. Further, even though the Black boy is clearly around twelve years old, the caption calls him a "man," implying he is older and more mature than he is. Some researchers call this *adultification*, and we see it often in how people talk about Black and brown kids, characterizing them as older, more responsible for their actions, and potentially more dangerous.

This "looting versus finding" example is *very* representative of what media research tells us is typical. When someone's behavior could be characterized as either more innocent or more criminal, White people are much more often given the innocent wording, whereas Black and brown folks much more often get the criminal wording. Across our lifetimes, therefore, our brains receive a biased pattern of input, which means we learn to think of people in more biased ways.

Although these Hurricane Katrina examples are over twenty years old now, we continue to see the same patterns in the media today. While I was writing this book, similar headlines popped up related to wildfires in Los Angeles, with Black families being characterized as "looting" when they were in fact trying to save their own possessions from their own homes. Recent work in artificial intelligence (AI) large language models provides even more evidence of these patterns. Like human brains, AI models get trained on media content, and then the AI models display the same types of biases we observe in human behavior. **Our cultural input teaches us stereotypes and biases that become our mental habits.**

THE PERSONAL DILEMMA OF UNINTENTIONAL BIAS

When we are adults, these bias habits are often at odds with our conscious values and intentions, but they nevertheless influence our behavior. In culture, there are many different labels for this phenomenon. Perhaps you've heard of *implicit bias* or *unconscious bias* or *automatic bias*; for the purposes of our journey in this book, those all refer to essentially the same thing. My colleagues and I prefer the term ***unintentional bias***, because it best captures the experience and reality of most people—when this kind of bias occurs, it's usually not intended. The vast majority of people are not *trying* to look down on someone or cause them problems because of their race, gender, sexual orientation, or other group status. Nevertheless, stereotypes pop to mind and influence one's reactions, judgments, and behavior toward members of stereotyped groups. This disconnect between automatic reactions and conscious values is what my colleagues and I call *the personal dilemma of unintentional bias*.

One example of this personal dilemma was shared with me by an attendee of my in-person trainings. Let's call her Jennifer. Jennifer is a feminist who cares deeply about fighting for gender equality and opposing sexism and gender bias. Those values characterize who she is as a person and what her conscious intentions are as they relate to gender. However, one day she fell short of these intentions at a work party. One of her colleagues wanted to introduce Jennifer to a visiting medical researcher, Dr. Smith. When her colleague brought her over to a man and woman, Jennifer enthusiastically shook the man's hand and said, "You must be Dr. Smith!" Then the woman next to him chimed in, saying, "Actually, *I'm* Dr. Smith." Jennifer felt so embarrassed she started apologizing profusely and compounded the issue by making a bit of a scene. She mentally beat herself up about this incident for a while, wishing she hadn't made the assumption in the first place, and also wishing that afterward she had just briefly apologized and moved on without making such a big fuss.

Jennifer's story illustrates how **unintentional bias can be a personal dilemma**. Remember, combating gender bias was personally important to her, but she still fell prey to the easily supplied stereotypic assumption that the man was "probably" the doctor rather than the woman. For Jennifer, it's a personal dilemma because she fell short of her own values and intentions, which say that she should not make stereotypic assumptions based on gender.

American civil rights leader Jesse Jackson once shared his experience with unintentional biases, saying, "There is nothing more painful to me at this stage in my life than to walk down the street and hear footsteps and start thinking about robbery. Then I look around and see someone White and feel relieved." Jackson is not alone in this experience. Many people, of all racial groups, report that when they notice a Black man walking toward them on the street, they often feel nervous or afraid, and may even cross the street to avoid walking past him, but wouldn't feel or act that way if it were a White man. These reactions are another example of unintentional bias, and they are the direct consequence of the cultural associations between race and crime we learned about earlier in this chapter.

These examples showcase key elements of the personal dilemma of unintentional bias. Despite conscious values that oppose bias, both Jennifer and Jesse Jackson still experienced an automatic, unintentional reaction that reflected stereotypic assumptions. These reactions are bias habits, and they can operate at odds with conscious values and intentions. Another important feature of these stories to note is that both involved people whose unintentional bias disfavored their own group—Jackson against fellow Black folks, Jennifer against women. Sometimes in modern discourse, people have the misperception that bias only happens from White people against people of color, or from men against women, or from straight people to LGBTQ+ people, and so on. But the reality is that *anyone* who grows up in a culture that has stereotypes and biases built into it will learn these bias habits and be vulnerable to expressing unintentional bias, regardless of their own identities. We're all human, and human brains learn biases from their social environments. The question then becomes, How do we break these bias habits?

APPROACH BIASES AS HABITS TO BE BROKEN

Think about other kinds of habits you've tried to break. Maybe you've tried to stop biting your nails, reduce your screen time, or get more exercise. The science behind changing these kinds of behavioral habits gives us an excellent model that we'll use in our efforts to disrupt bias habits. Breaking any habits, including bias habits, requires Motivation, Awareness, Tools, and Effort.

Motivation

To make any kind of change, you have to decide that you care about changing the habit. You need ***Motivation***. No one else can make that decision for you. Working to disrupt bias habits has to be something *you* think is important and something you're committed to doing.

Someone I will call Trina used this model in her efforts to break her lifelong habit of biting her nails. When she was young, her mother would try to get her to break that habit by scolding her, saying things like, "You're getting dirt in your mouth" or "You don't look like a proper young lady." As many kids do when their parents scold them, Trina dismissed and ignored her mother's comments. Someone else trying to push you to break a habit is a very poor way to create change. Trina did eventually stop biting her nails, but only once she reached a point in her life where it was what *she* decided for herself. Only when she was herself motivated to change was she able to break the habit.

The same principles apply to bias habits. Each of us needs to make a personal decision that reducing the potential for bias is something we personally care about and want to work on. Change doesn't happen because I explain to you all my reasons for thinking you should work on reducing bias—it has to come from within you. But there's good news! As I mentioned earlier, *most* people—truly, the vast majority—oppose bias in their own behavior. People want to treat others fairly, unencumbered by race bias, gender bias, or bias related to other identities. When people come to understand that they could be vulnerable to expressing bias without realizing it, they are indeed *motivated* to break those bias habits.

Awareness

The second key component of our model for change is ***Awareness*** of how the habit plays out. You need to understand how the habit manifests in everyday life: what it feels like, what it looks like, what initiates the habit, and what keeps the habit happening over time.

Trina's nail-biting story also dovetails nicely with this awareness component. Trina didn't even realize when she bit her nails. She just knew that when she looked at them, they were all nibbled. A key feature of habits is that we aren't always aware of them; they usually happen on autopilot. Once Trina decided she wanted to change (once she was motivated), she started tuning in to her behavior more carefully, figuring out when it was she bit her nails. Was it when she was waiting for a meeting to start? When she was bored? Hungry? Anxious? She had to identify what sorts of external or internal cues led to the habitual behavior, so she could plan on how to disrupt it.

In the case of bias, awareness skills involve tuning in to the different ways stereotypes and biases can influence your thoughts and behaviors. What does bias *feel* like? How do stereotypes in our minds produce discriminatory behaviors? What things happen in our minds that perpetuate bias? Bias can play out in many different ways, and we'll learn to recognize and root them out in our daily lives. These awareness skills will be the focus of the chapters in unit 2 (coming up after this chapter), which will help us learn about bias habits so we can disrupt them.

Tools

Once we've tuned in to how bias plays out, we need effective ***Tools*** to break out of the old habits and develop new habits that are more in line with our values. Breaking a habit is a process of *both* reducing the old behavior and building up new behaviors.

If the habit is nail biting, you might try squeezing a stress ball, wearing gloves, sitting on your hands, or applying a bitter-tasting nail polish. The tool you use might depend on what you learn in the awareness step. If the habit is triggered by anxiety or having too much fidgety energy, the stress ball might be most suitable. For someone who bites their nails in their sleep, going to sleep with gloves on might be the best tool.

While a stress ball might be sufficient for nail biting, addressing bias habits is a bit more involved. We're going to learn a variety of tools to help us disrupt bias habits from many different angles. Some of the tools will be mental exercises that help reduce bias in our minds, whereas others will be more interpersonal and external, involving actions we can take to create more inclusive environments, thereby cultivating more diverse joy. Taken together, the tools we learn about in unit 3 will equip us to tackle the various bias habits we will learn about in the awareness unit.

Effort

Lastly and importantly, no habit gets broken overnight. It takes ***Effort*** over time to disrupt well-learned habits. In other words, you have to actually put the awareness and tools into practice.

Remember that bias habits are ingrained patterns that have been reinforced for your entire life. As adults, we continue to be bombarded with biased and stereotypical input from the media and our social environments, an extension of how, as kids, we first learned biases. We need to put in effort over time to counteract what is already in our minds and what we continue to be exposed to, and to put into practice our new ways of responding, thinking, and behaving. Remember that breaking a habit is a process and you will slip up from time to time. Reading this book won't erase all biases from your life. In fact, you'll start noticing and recognizing *more* instances of bias as you develop the bias habit-breaking skills. There may also be some trial-and-error as you figure out the best way to apply these skills to combat bias in your life. But putting in effort over time will help you wear down the unintentional bias habits and build new behaviors that are more in line with your conscious values and intentions. Practice makes progress.

How to Create Meaningful Change

Motivation, Awareness, Tools, and Effort are the four components that make up the core model of change to keep in mind as we work on disrupting bias habits. The Awareness and Tools components are the key

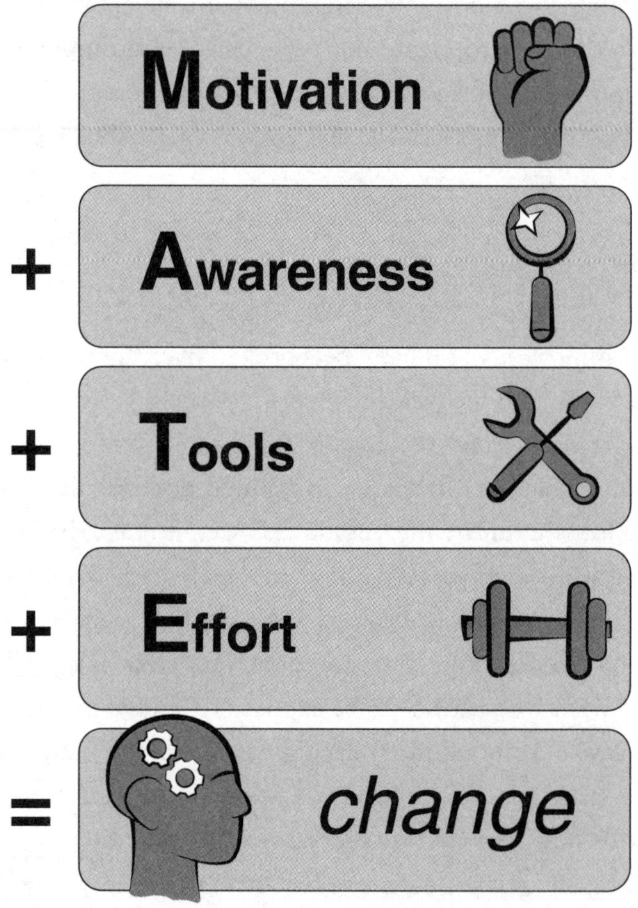

Figure 2. Our model for creating meaningful change.

skill areas that this book can help you develop, but the Motivation and Effort have to come from you. If you're reading this book, chances are you're already motivated to work on overcoming bias, and learning more will only increase that motivation! The activity of reading this book and reflecting on its content is also the first bit of effort over time, getting you started on this journey.

This habit model makes up the **core strategy** for how we'll learn to disrupt bias habits on this journey. Whenever you discover or notice a new form of bias or a bias habit you want to break, you can come back to these four components as you think through your plan of action for how to deal with those biases moving forward. For example, a few years back I used this model to break a verbal habit that was pointed out to me. When speaking to a group of people, I would say "you guys" when referring to the collective group, regardless of the gender makeup of the audience. If I was addressing my students, for example, I'd say, "Hey, you guys, open your books to page 27." One day, a friend named Phoebe pointed out to me that it bothered her when people used that phrase, because it implies that everyone in the group is a guy. It also bothered her in particular because growing up she was a bit of a tomboy, and other kids bullied her for that, saying things like, "You're not a girl—you're a *guy*!" Because of those childhood experiences, whenever someone used the phrase "you guys" to refer to a group she was in, it would give her an emotional jolt as she remembered that bullying and nervously wondered if the speaker was insulting her.

Because Phoebe pointed out that verbal habit, I thought about it and decided I wanted to change it. So I did the following:

- Using our model, I started by thinking about my *motivation* to break the habit. Why did I want to stop saying "you guys"? First and foremost, I'm the sort of person who cares about using inclusive language, and as someone whose *career* is to help people with bias and inclusion, I wanted to hold myself to high standards for my own speech and behavior. Also, having learned how that phrase had a negative effect for Phoebe in particular, I wanted to change it so I didn't make her or others feel bad.
- Having shored up my motivation, I looked to enhance my *awareness* of this verbal habit. I knew I said "you guys" often, but I began to monitor my speech carefully to notice when I said it and try to disrupt it as it happened. I found that I used that phrase *most* often when talking to my students, so in addition to

generally monitoring my speech, I paid special attention when in a class or lab setting when I'd be talking with students.

- Then, I needed *tools* to help me change. One of the tools I'll cover later is to "rehearse replacements," where you practice a new response to replace the habit. In this case, I decided to start saying "you all" or "y'all" to replace "you guys."
- Then I put in *effort* over time, trying to modify my speech. Sometimes I would slip up and then correct myself, so if I said, "Hey, you guys," I'd catch myself and follow up with "I mean, you all." As I practiced, those slipups became less common, and today my verbal habit has fully changed to saying "you all" every time. Now, I only ever say "you guys" when I tell this story!

My "you guys" example demonstrates how **we can always come back to the four components to address a new problem**. We're all lifelong learners and can discover new bias habits we might want to change at any time.

Values and Context Matter

This story also helps emphasize a central tenet of this training: although I shared my thought process behind changing my own behavior, I didn't demand that *you* should stop saying "you guys," nor did I imply that you're naughty or biased or sexist if you use that phrase. This book is not a list of things my personal values say you should and shouldn't do—if it were, this would be a book of opinions and personal morality, not science. As noted earlier, this training is about *skills*, and how you apply those skills is up to you! I'll share many examples of how myself and others have used these skills, but **it's up to you to decide what's right in your own life and behavior**. I hope you'll thoughtfully consider the examples I share, but I'm not writing this book to dictate your morality.

I've used the "you guys" example in countless presentations, and it also gives us a nice case study to **consider context**. For audiences in science fields, many women report having similar feelings to those of Phoebe, because of the historic underrepresentation of women in most science fields. Many people in science identify man-centered language as an everyday form of bias that creates a less inclusive climate for anyone who doesn't identify as a "guy." In science contexts, therefore, I'd infer that more people would want to change that particular verbal habit.

In other contexts, however, things may be different. When I used the "you guys" example with an audience of police officers, several women officers shared their differing perspective. For them, a big part of their career path was figuratively becoming "one of the guys," through the physical training and boot camps all officers have to complete. For those women, being "one of the guys" was a badge of honor, and when people said "you guys" to a group including them, it filled them with a sense of pride. Why would I want to rob them of that feeling of pride and connection to their colleagues? I wouldn't! This example, like all the examples in this book, are *just* examples. We're using these examples to learn new skills. I invite you to *consider* whether the way others have used those skills might be relevant for you. But only you can decide what's right for your own behavior.

There are very few hard-and-fast rules in the realm of bias and diversity topics. The only one I'd suggest is **"Don't say slurs."** For everything else, there's substantial context, nuance, and personal discretion to consider.

ADOPT A CUMULATIVE PERSPECTIVE ON BIAS

I'd like to end this chapter with an important discussion of how bias impacts others. The examples of bias I've shared thus far vary in the extent to which their impact is readily apparent for everyone.

Assuming someone is a criminal because of their skin color, for example, is something most people would agree is morally wrong and potentially very harmful. Other examples, however, may seem relatively minor, especially from the outside. After all, is it extremely harmful to assume a man is more likely to be a doctor than a woman or to say "you guys" when referring to a group of people who aren't all men? It's not always clear to everyone why these kinds of examples matter, but they tend to matter a lot to the people who are the target of these types of bias, because they have cumulative impacts.

If I slip up and say "you guys" to a group once, my slipup is probably *not* the only time that day or week that the non-guys in the group have heard gendered language that doesn't include them. These individual incidents are not isolated experiences; they add up and have a cumulative effect over time. Many people who are members of groups targeted by bias say that a single instance of bias may not be catastrophic when considered *in isolation*, but these instances of bias happen frequently, and the **accumulation of many bias incidents over time is exhausting and can be overwhelming**.

Imagine that Eric is a White person who, the majority of the time, doesn't express bias when he encounters Black people. After all, just because people are *vulnerable* to bias habits doesn't mean they always display bias. From Eric's perspective, bias seems rare and isolated, because most of the time he behaves fairly. Within the broader context, however, White people outnumber Black people in the US by a large margin. This imbalance means that even if most White people are like Eric and very infrequently express bias, most Black people end up experiencing bias constantly.

I've emphasized how bias has a cumulative impact on how someone feels in the moment, but I also want us to keep in mind the cumulative impact on someone's life over time. Bias incidents add up over days, weeks, and years and affect people's mental and physical health and

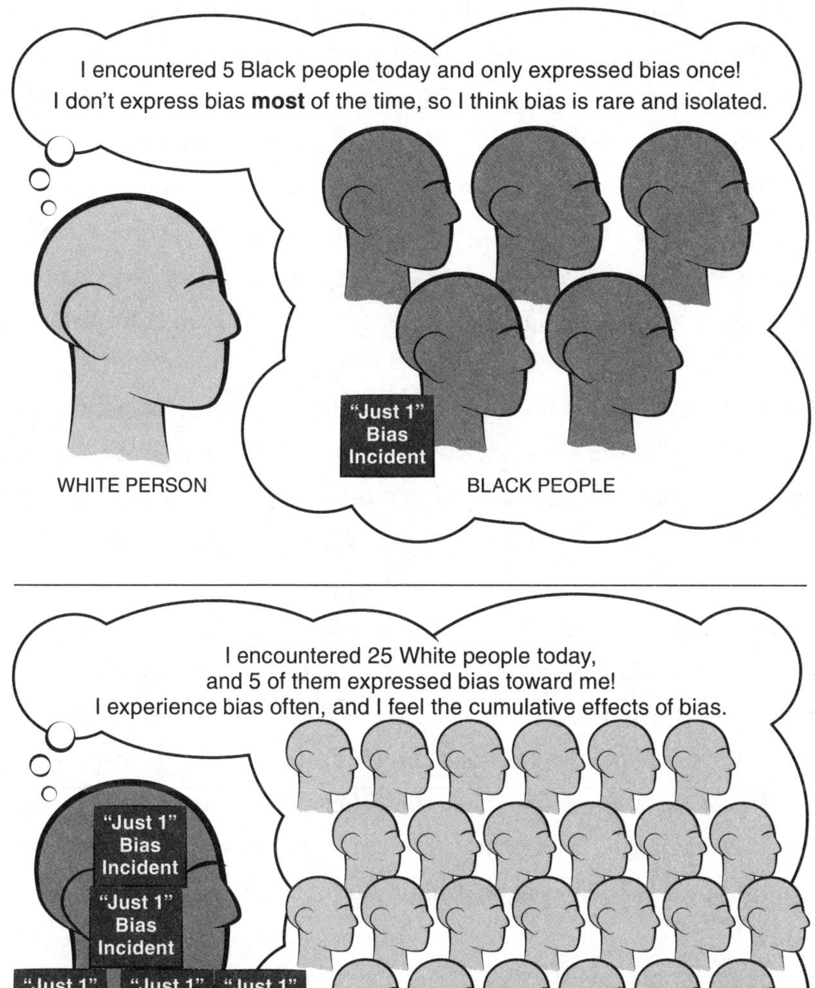

Figure 3. Two perspectives on the same instance of bias.

well-being. People get burned out and change careers because of the cumulative onslaught of many ostensibly "small-in-isolation" incidents of bias. While writing this, I asked a doctor friend of mine whether it would be accurate to say that she has patients question her competence because she's a woman "at least once a day," and she just started laughing, because it happens to her and her women colleagues so much more often than once a day! Mitigating our contributions to these cumulative impacts is part of why we want to work on disrupting bias habits.

Understanding how bias has a cumulative impact is especially important when we're the ones who unintentionally express bias. Often our knee-jerk reaction is that we want to forgive ourselves for making a mistake, we want to keep thinking of ourselves as good people, and we therefore tend to try to minimize the impact our mistake has on someone else. We also keenly remember all the many other times we *didn't* express any bias! All of this encourages us to see the bias incident as isolated and not a big deal. It is essential to remember that **you're not a bad person because you made a mistake, but the mistake still likely does some harm** to the other person. We're much better served by trying to understand that cumulative impact, both to make genuine amends and to help maintain our motivation to do better in the future.

One final note on intensity of impact. Sometimes in this book I'll use a seemingly mundane example because it makes the skill we're learning easy to understand, or because I want to avoid the intense emotional impact of a more extreme example. But I want you to keep in mind that even examples that seem minor in isolation can add up and have a bigger, cumulative impact on the people targeted by bias. Also remember that any example I use is just a way for us to learn about the fundamental processes that give rise to biases, but those biases could have a variety of levels of intensity.

CHAPTER 1 SKILLS SUMMARY

I want to be explicit about the actionable skills from this chapter.

- Having learned how cultural input teaches bias habits, start to **Monitor Your Input**, noticing when and how media and other aspects of your social environment might expose you to stereotypical and biased ideas.
- Now that you've learned about the personal dilemma of unintentional bias, **Recognize That Bias Can Happen Unintentionally**. As we continue this journey together in the next three chapters, we'll learn more specific details about the ways bias habits play out!
- Keep in mind that we're trying to **Approach Biases as Habits to Be Broken**. Like other habits, bias habits can be broken, if we're motivated to change them, are aware of how they operate, have tools to build new behaviors, and put in effort over time. (If you like acronyms, the four components of our model of change—Motivation, Awareness, Tools, and Effort—happen to spell out the word *MATE*. That wasn't intentional, but you can use it if you find it helpful!)
- Lastly and importantly, **Adopt a Cumulative Perspective on Bias**. Especially when you're the one who has accidentally expressed bias, you may have a tendency to see that bias as a rare, isolated incident. But remember that for the person who is the target of bias, it probably isn't a rare occurrence, and what seems small to you may have a big, cumulative impact on them.

REFLECTION ITEMS

- Can you think of times you've seen stereotyped or biased portrayals in the media? How do you think they might affect you?

- Can you think of a time from your childhood when you expressed bias or learned a stereotype? Or a time another child in your life did? What might some of the consequences of that stereotype or bias be?
- Have you ever experienced the personal dilemma of unintentional bias, where an automatic reaction conflicted with your conscious values? How did you respond?
- Think of some occasions where you witnessed, heard about, expressed, or were targeted by bias. Focus especially on instances of *unintentional* bias (e.g., stereotypic assumptions about someone). What might have been some of the harm caused by this bias, and how could it have been avoided or addressed better?
- What are your initial thoughts on our model of change, which involves Motivation, Awareness, Tools, and Effort? Which component do you think will be most challenging for you? Try to generate an example or two that uses the four components to break a habit.
- Can you think of a time when something that might seem small in isolation had a big, cumulative impact on you or someone else? (This could relate to bias or to something else.)
- See if you can generate examples that illustrate how this chapter's content is
  - *actionable* (What are you *doing*?)
  - *self-sustainable* (How will you maintain it over time?)
  - *generalizable* (Consider examples related to race, gender, sexual orientation, age, disability, politics, religion, body size, or other groups.)
  - *customizable* (How might you use these skills in your various work, life, and play contexts?)

SHARE YOUR REFLECTIONS!

Your examples and experiences can contribute to our research, help us improve this training program, and provide guidance to others who complete this training in the future! Consider sharing your examples, insights, and ponderings with us at BiasHabit.com/share.

UNIT TWO

AWARENESS SKILLS

Bias Habits to Disrupt

THIS UNIT WILL HELP US develop our awareness skills. Specifically, in chapter 2, we'll learn some fundamentals about how stereotypes in our mind influence our expectations, emotional reactions, and behavior. Then we'll use those fundamentals as building blocks to understand several different bias habits in chapters 3 and 4.

People often experience bias as some vague sense that something is wrong. We know "something doesn't quite feel right" or "this seems like some kind of bias" without being able to fully articulate it. The awareness skills in this unit will help us move beyond that vague feeling to **develop a step-by-step understanding of how bias habits operate** and how to **tune in to our thoughts and emotional reactions** in order to disrupt those bias habits. Understanding *how* stereotypes and biases give rise to unfair feelings, judgments, and behaviors is key to our ability to break bias habits.

2

BIAS HABIT BUILDING BLOCKS

Expectations and Emotions

IN THIS CHAPTER, we're going to learn a few fundamentals about how our minds and brains work that serve as the building blocks for bias habits. Each of the sections in this chapter teaches us about a crucial idea or feature of our mental processes that will be essential in understanding the bias habits we explore in the next two chapters and the tools to disrupt them that we'll learn about in unit 3.

STEREOTYPES GUIDE EXPECTATIONS

Once a stereotype has become habitual in our minds, how does it operate? How do stereotypes lead to biased assumptions, judgments, and actions? When we encounter or think about someone who is a member of a stereotyped group, it activates the stereotypes associated with that group, guiding our expectations. For example, if we see a Black man, the stereotype associating Black men with criminality will be activated; thus we might unfairly expect the Black man could be criminal and dangerous.

Stereotypes we've learned may lead us to expect that women are more likely to be homemakers than doctors, that people from Wisconsin love cheese, that women with very short hair are likely to be lesbians, that Asian people are good at math, that men aren't good with children, and so on.

Some neuroscientists argue that **prediction is the primary purpose of our cognitive systems**—our brains are constantly generating expectations about what will happen next to help us successfully navigate the world. These expectations can include whether someone will be friendly or hostile toward us, whether a situation will be safe or dangerous, or whether a food will be tasty or disgusting. For everything in our lives, the predictions our brains generate arise from the patterns in our past experiences. Importantly for bias habits, our past experiences include *both* the people and events we have actually encountered *and* representations from the media and messages in our social environments that do not involve actual encounters. **These experiences set expectations that may not reflect reality.**

Depending on the circumstance, these expectations may have dire consequences or very mundane ones. An example of the former involved stereotypes about Indigenous people and a colleague of mine, a medical doctor whom I'll call Dr. Ruby. In the US and Canada, stereotypes of Indigenous people (aka Native Americans, American Indians, or First Nations people) say that they are especially likely to have problems with drunkenness and alcoholism. Ruby and her sister, Mavis, are Indigenous people. One day, Mavis had a seizure, and Ruby took her to the emergency room (ER). While Mavis's condition was assessed and triaged, Ruby was in another room filling out some paperwork. The ER staff, upon seeing Mavis's slurred speech and difficulty standing up, determined that she was drunk and started to send her to the sobriety center, where she could spend the rest of the night sobering up. Luckily for Mavis, Ruby was present and corrected this error, explaining that not only was Mavis sober on this occasion, she actually *never* drank alcohol. Period. If Ruby hadn't been there, rather than Mavis getting the appropriate treatment for her seizure, she would have sat in the sobriety center overnight, compounding the damage to her brain. The stereotype of Indigenous people as drunks created an expectation that led the medical personnel to think Mavis's slurred speech meant she was drunk, a quick but catastrophically incorrect assessment that could have seriously harmed her.

Not every case of stereotypes guiding expectations is as extreme as in Mavis's example, but stereotypes can guide our expectations in many ways. And **these expectations may be fully conscious or fully unconscious**. Imagine that I'm talking to my friend John, who is a White man from the US and of average height. We hear a knock on the door and I tell John, "Oh! That must be my friend Lamar!" When John turns to greet Lamar, his head tilts slightly upward, because his tacit expectation is that Lamar will be tall. This isn't a conscious process, by which John thinks to himself, "Hmm, this person is named 'Lamar,' so he's probably Black. And if he's Black, that probably means he's tall, so I better tilt my head upward so I can make eye contact." Odds are, *none* of that went through John's mind consciously. But nevertheless, his automatic associations—his mental habits—did lead to that chain of expectations (Lamar → Black man → tall → tilt head up).

People tend to think of Black men as especially tall, but the data say that if there's any reliable height difference between Black men and White men in the US, it's that Black men on average are actually a little *shorter* than White men. We just tend to imagine Black men as taller on average because of the prominence of tall Black basketball players in the media. This ties back to chapter 1—our mental habits and stereotypes come from the *input* our brains receive, which is not necessarily an unbiased accounting of real-world data. Generally, it is the ideas with the most repetitions, not the ones with the strongest basis in truth, that exert the biggest influence on our expectations.

STEREOTYPIC EXPECTATIONS ARE PROBABILISTICALLY UNCERTAIN

A *stereotype* is **a learned association between two human concepts that do not define one another** (e.g., *Black–criminal*). When looking at group-level similarities and differences, stereotypes are not inherently accurate or inaccurate by definition. Sometimes, a stereotype may converge with or reflect a real correlation in the world. For example, in the US, the *Black–poor* stereotype corresponds to a real correlation between race and socioeconomic status, due to pervasive historical and contemporary racial inequities. Other times, the stereotype may directly contradict the real-world data, like in my earlier example about how the *Black–tall* stereotype doesn't match the data showing that Black men are shorter than White men on average. Regardless of whether there is a statistical relationship at the group level, however, there are no stereotypes that hold true for every member of a social group. When someone is standing right in front of you, you really have no idea whether stereotypes will actually apply to them as an individual. Some members of stereotyped groups will happen to match a stereotype (e.g., a Black man who likes basketball), and some people will not (e.g., a woman who prioritizes career over family). In other words, **stereotypic expectations are probabilistically uncertain**.

Even when a stereotype happens to correspond to a real group-level statistic (e.g., the *Black–poor* stereotype in the US), most people tend to agree that judging a single person based on that group-level correlation (e.g., assuming "*this* Black person in front of me is probably poor") is morally wrong. Aside from that moral argument, using a stereotype to make an assumption about an individual is also unlikely to result in an accurate judgment. The group-level correlation could only hope to be informative if we had a truly random statistical sample. For instance, if you randomly selected one Black person and one White person from the US, it would be slightly more likely that the Black person would be poorer. But even then, you would also have many cases where the White person is poorer, because they come from heavily overlapping statistical distributions. More importantly, our lives do not involve random sampling! Odds are, more of the people you regularly interact with will be close to your own income level, due to the neighborhood you live in, whom you have as friends, and your coworkers likely having similar levels of education to you. People are complex, and demographic group identities are just one part of who they are. Relying on a group stereotype to reach a conclusion about an individual always has high probabilistic uncertainty.

CONFIRMATION IS SATISFYING

Imagine that you roll a six-sided die and guess that it will land on a three, and then it does! Despite your receiving no external reward, the correct guess feels satisfying—you feel lucky. Your brain likes that your uncertain prediction was upheld. The same thing happens with stereotypic expectations: when they happen to be confirmed, it engages your brain's reward system and "feels good"—even if it was just random luck that the stereotype happened to be confirmed. When a probabilistically uncertain expectation is upheld, it is satisfying—it feels good, and we like it. In fact, your brain's reward system is *especially* active in cases of high probabilistic uncertainty, like stereotypic expectations. Now, imagine rolling a six-sided die that has been weighted so it lands on six most of the time. You predict six, then roll a six, predict six, then roll a six, and so on. It's not as much fun anymore, is it? That's because it has become too predictable. The uncertainty is an important part of this reward involvement.

If you've ever taken an introductory psychology class, you might remember the term *intermittent reinforcement*, which essentially refers to a pattern of expectation confirmation that is probabilistically uncertain. In old research studies, rats would learn to push a button to receive a treat. One group of rats would get a treat every time they pushed the button, so the button had a high degree of predictability, with *no* uncertainty. Another group of rats would only get a treat intermittently—maybe one out of every ten times they pushed the button. For these rats, therefore, the button had very high probabilistic uncertainty. In the next phase of the study, the scientists made it so that the button would *never* give any rat a treat. The rats in the first group quickly learned that the button "no longer works" and stopped pressing it. The rats that had the uncertainty, however, kept coming back to the button and essentially never learned that it didn't work. These and many other studies demonstrated that **the probabilistically uncertain reward pattern is the hardest learning pattern to break**.

In human behavior, we see this same learning pattern perhaps most readily in things like gambling addiction. People know that the lottery or slot machines are very probabilistically uncertain (even if they don't use that term), but on the rare occasions when the slot machine pays out, the neural reward is so strong it often become addictive. (With this example, I'm adopting a simplified view to help illustrate this section's main point; there are, of course, many additional social, cognitive, and neurological factors at play in actual gambling addiction.)

Returning to stereotyping and bias, just like pulling on a slot machine or rolling a fair die, most of the time your stereotypic expectation won't be confirmed. You pass someone on the street, have a quick expectation or assumption about them, and never actually learn whether your expectation was correct or incorrect. On rare occasions, however, you'll get confirmation that your assumption was correct, which will activate your brain's reward system, leading to a pattern of learning that is extremely hard to overcome.

I want you to start tuning in to how this reward learning might feel for you. Consider the following example. In the US, many people have the stereotypic expectation that bad drivers are women. Imagine a man who is a bit sexist and holds this stereotype in his mind. When someone in another car cuts him off, he shouts something like, "Geez! Watch it, lady!" or "Probably a woman driver!" When he passes the other car, he sees that it is in fact a woman driving and says something like, "Haha! I knew it!" The "I knew it!" reaction is often how this satisfaction manifests for people in response to stereotype confirmation. The man in the story felt satisfied that his prediction was correct.

I set up this story with the idea that this man is a bit sexist, but in my many years of using this example, I've had a great many people, including many women, affirm that they have that sort of "I knew it!" reaction about women drivers and then felt guilty afterward. Your brain is going to like it when uncertain expectations are confirmed, even if, morally, you don't agree with the expectation. That's where the guilty feeling comes from; after you have the reward reaction, you realize it's biased in a way you don't approve of and perhaps feel some guilt. In a situation like this, the guilt is useful because it helps you identify your values and the way your initial reaction violates those values.

Just in case you were wondering, US data say that men are worse drivers than women on average, by almost any objective metric (e.g., being the cause of accidents, getting speeding tickets, driving without a license).

A journalist once asked me about this reward/stereotype confirmation connection and specifically wondered about stereotype confirmation related to an objectively unpleasant outcome (e.g., someone being a dangerous criminal is inherently scary, so why would we "feel good" about that stereotype being confirmed?). The analogy she used was to think about having a father who has a reputation for giving terrible presents. You predict that his present will probably be something you don't want, then when you open it, sure enough, it's terrible. Your uncertain expectation was confirmed, and now you are stuck with a terrible present. Shouldn't the bad present lead to negative feelings, regardless of whether it confirmed an expectation? In response to her, I first acknowledged that, of course, **mixed emotions are a part of life**!

We can feel more than one way about something! I then asked her whether she could imagine joking with her spouse about the bad present. Might she say something like, "Can you believe it? I knew my dad would come up with something awful like this! Haha!" and have a little laugh about it? The laughter and amusement comes from the satisfaction of the expectation being confirmed.

Just like a driver might both feel that "I knew it!" satisfaction when a bad driver is a woman but then feel guilty about having a sexist thought, the journalist can laugh at her prediction being upheld and still be unhappy with her present. Further, all the things I discuss in this book can **operate at different levels of intensity and consciousness**. Some of these reward-related feelings may be very conscious, and some will happen below conscious awareness. But learning about them will help us tune in to how the bias habits play out, so that we can disrupt them.

DISRUPTION IS FRUSTRATING

Now that we've learned about how expectation confirmation is satisfying, what happens when our expectations are *dis*confirmed? When our expectations turn out to be wrong, or they get disrupted, it is aversive; it feels bad. This aversion often manifests as a feeling of irritation, annoyance, or frustration. We get these emotional reactions with all different sorts of habits when they get disrupted. Have you ever gone to a grocery store that's laid out differently than the one you're used to? It's mildly annoying when the bread isn't where you expect it to be. These feelings are the direct result of expectation disconfirmation or habit disruption. You're used to the easy, typical, habitual way of doing things, and the disruption results in irritated feelings. This pattern happens for all sorts of habits—bias habits included.

Recall the man in the "bad driver" scenario from the previous section, but imagine that he discovered the person who cut him off was actually a man. In addition to any frustration he felt for having to deal with the bad driver, there would be added annoyance because of his expectation being disrupted.

Suppose you're a person who habitually walks at a fast pace, but today you're going somewhere with a friend or a colleague who uses a wheelchair or has to walk at a slower pace because of an injury or a disability. Even though you have plenty of time to reach your destination at the slower pace, the disruption to your fast walking habit is going to give rise to some frustrated or annoyed feelings. We often don't like to label these experiences as *annoying*, because it's rude to call a *person* annoying. But we need to recognize that these **frustrated emotional reactions are the natural consequence of our habits being disrupted**. We can think of similar examples, for instance, with talking. Imagine that someone who is used to talking at a rapid rate has to interact with someone who's using a translator or a speech assistance device, which makes the conversation necessarily go slower. Again, the disruption in the habitual way of talking will bring up frustrated and annoyed feelings.

These emotional reactions can also arise from disruptions to our habitual ways of seeing or thinking about the world. For instance, when lesbian, gay, and bisexual people started coming out and becoming more prominent in US culture, many people reacted with hostility, in part because it disrupted their habitual way of thinking of everyone as straight. Heterosexuality was the norm. Something different from what we're used to, our worldview, or our thoughts on a topic is another kind of mental habit disruption. We've seen this more recently with the increased recognition of transgender people and people with gender nonbinary identities in culture and in the media. Although we know from historical records that nonbinary people and transgender people have existed throughout human history, for many people, these identities are a "new idea" to which they've only recently been exposed. The disruption in their mental habit of how they think about gender and gender identity will bring up frustrated and irritated emotional reactions. We want to understand how this disruption creates frustration because it ends up feeding into many of the bias habits that we're going to learn about.

Consider another example. Suppose John is telling his friend Nick about a plumbing problem, and he says that the plumber he got was really wonderful. Nick also needs a plumber and asks, "What's his name?" John replies, "Well, *her* name is Jill!" How does Nick feel? He probably feels slightly annoyed or frustrated, even if only a little bit. Again, his expectation that the plumber would be male wasn't the result of a conscious deliberation in which he thought, "Well, most plumbers are men, especially the great ones, so *surely* this one has to be a man." The expectation was more automatic, or tacit, the result of a mental habit. Some of the annoyed feeling might come from Nick thinking John is accusing him of being sexist, but at least part of it arises from his habitual expectation being disrupted or disconfirmed. If I were in Nick's position, I would have felt the mild annoyance at my habit being disrupted, then some added annoyance at myself, because avoiding biased assumptions is important to me, so I fell short of my own standards. I would then feel happy or excited about the woman plumber, because I tend to be eager to support women in historically man-dominated fields. Don't forget, mixed emotions are a thing!

The takeaway here is to remember that **habit disruption or expectation disconfirmation will bring on frustrated or irritated emotional reactions**. They may be very low-level or very overt, and they may come mixed with other thoughts and feelings. But I want you to understand the broader point that habit change, by definition, will involve some discomfort, and the two more specific points that your brain "likes" expectation confirmation and "dislikes" expectation disconfirmation. These two specific points will be foundational as we continue learning ways bias plays out in our judgments and behavior.

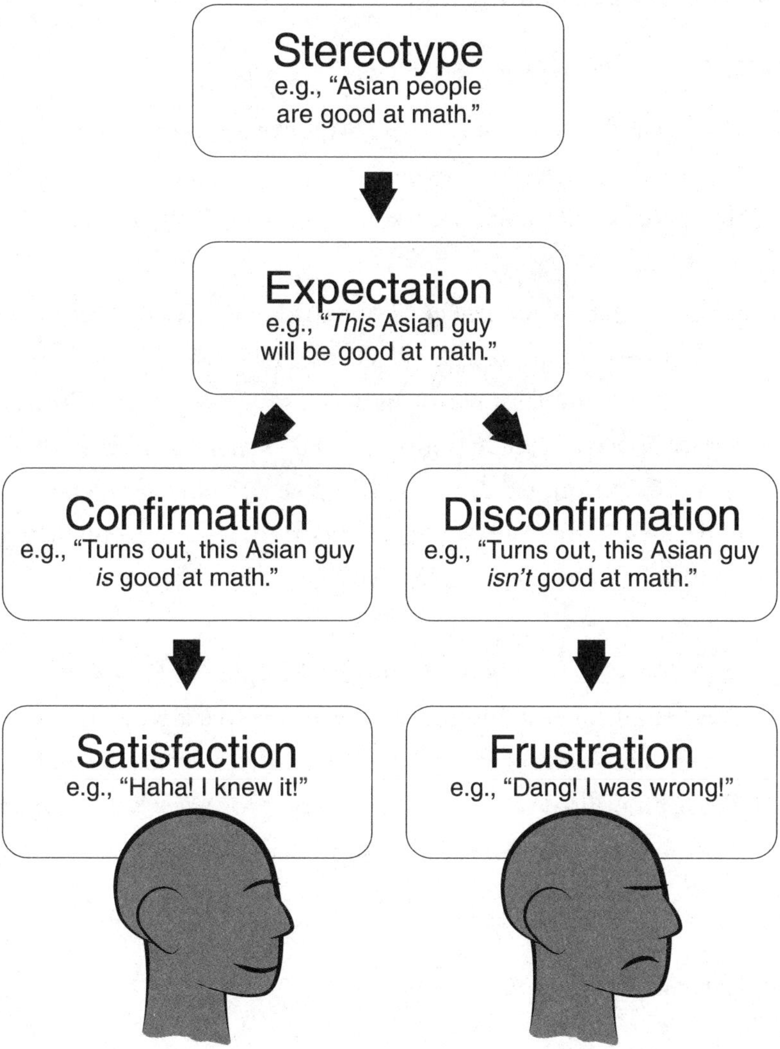

Stereotype
e.g., "Asian people are good at math."

Expectation
e.g., "*This* Asian guy will be good at math."

Confirmation
e.g., "Turns out, this Asian guy *is* good at math."

Disconfirmation
e.g., "Turns out, this Asian guy *isn't* good at math."

Satisfaction
e.g., "Haha! I knew it!"

Frustration
e.g., "Dang! I was wrong!"

Figure 4. Stereotypes guide expectations, which result in emotional reactions.

CHAPTER 2 SKILLS SUMMARY

Taking what you've learned in this chapter,

- Try to start to **Recognize How Stereotypes Color Our Expectations**. Stereotypes can lead us to expect or anticipate that people will match the stereotypes about their group identities.
- Understand how **Stereotypes Are Probabilistically Uncertain**. Stereotypes about groups, even those that may have some real correlation with features of the social group as a whole, lead to highly uncertain predictions about any single member of that group we encounter. Also, most people don't like it when you make an assumption about them because of their group identities. More often, they'd rather you get to know them as an individual.
- Start trying to **Notice the Emotional Impact of Expectation Confirmation and Disconfirmation**. We like it when our expectations are upheld and dislike it when they are disrupted. These emotional reactions reflect habit processes, not necessarily what we consciously *believe* about the other person.

REFLECTION ITEMS

- What are some stereotypic expectations you have about social groups other than the examples used in this chapter?
- What are some expectations relevant to your work or life that could be biased by stereotypes?
- What are some stereotypes that might lead you to have an "I knew it!" reaction? Can you think of a time you or someone else "liked" it when they got stereotype confirmation?

- Think of a time you were annoyed when a habit was disrupted or an expectation was disconfirmed. How did you deal with it?
- See if you can generate examples that illustrate how this chapter's content is
  - *actionable* (What are you *doing*?)
  - *self-sustainable* (How will you maintain it over time?)
  - *generalizable* (Consider examples related to race, gender, sexual orientation, age, disability, politics, religion, body size, or other groups.)
  - *customizable* (How might you use these skills in your various work, life, and play contexts?)

SHARE YOUR REFLECTIONS!

Your examples and experiences can contribute to our research, help us improve this training program, and provide guidance to others who complete this training in the future! Consider sharing your examples, insights, and ponderings with us at BiasHabit.com/share.

3

BIAS HABITS THAT CREATE DISPARITIES

IN THIS CHAPTER, we'll see how the building blocks we learned in chapter 2 play out in a few different configurations to learn about three specific bias habits: impression justification, norm enforcement, and self-fulfilling prophecy. Each of these bias habits demonstrates a way that we get from the biases in our minds to actual biases in our behaviors toward, and judgments about, other people, in ways that create disparities for members of stereotyped groups.

IMPRESSION JUSTIFICATION

Our first bias habit in this chapter, **Impression Justification**, involves how the judgments we make can be influenced by the reward and aversion processes we just discussed. Similar to how our brains like confirmation, **our brains also like it when information is easy to process**. Some scientific terms for this are *perceptual fluency* or *processing fluency*. Just as it's easier for you to understand a language you're fluent in, it's easier for our brain to process information out in the world when it matches our preexisting expectations. This fluency or ease results in a "good gut feeling," arising in part from the reward processes we discussed in chapter 2. On the flip side, just like it's harder to understand a language you're less fluent in, our brains have to do more work to process information that mismatches our expectations. This disfluency results in a "bad gut feeling," as our brains have to work harder to figure things out, arising in part from the frustration we discussed earlier.

These good or bad gut feelings become the lenses through which we make judgments or decisions. To make this more concrete, imagine what a police chief looks like. For most people, the first image that comes to mind is a man, because we tend to think of police chiefs as being men. Our stereotypic expectation, therefore, is that a police chief will be a man.

Now imagine that we're tasked with picking a new police chief, and we look at a man's application for the job. He matches the expectation of a man in the police chief role—he's the square peg in the square hole of our mental habit—so we'll have a good gut feeling about him, since he's easier for our brain to imagine in that role. That good gut feeling is the starting point as we look at his résumé. He has five years of experience on the police force, and on top of that, he even worked with researchers and was an author on an academic paper showing how to measure the effectiveness of different police practices. All of that looks pretty good. We think something like, "Wow, look, at his great experience. And hey, he's published this paper. When he's police chief, and he enacts a new policy, people will really believe in it, because of this research he did! He's pretty great!" We started with the good gut feeling and then we interpreted everything positively through that lens.

But what happens if we instead imagine that a woman submitted that same application? The woman *mis*matches our expectation—she's a round peg in the square hole of our mental habit—so our brain has to work harder to picture her as the police chief. She had the same five years of experience, she'd published that same research paper, but because her being a woman doesn't quite "fit" our mental expectations, we start with a more negative gut feeling. We say something like, "Five years? Is that really enough experience? Oh, yeah, she's published a paper, but that's academic stuff, it's not *real* police work. She probably won't make a good police chief."

These gut feelings are our initial impressions of the two candidates, and then we interpret evidence in a way that *justifies* our *impressions*. Hence the term *impression justification*.

The candidates' "fit" with our stereotypic expectation led to different initial impressions about each of them, and then we built a story to justify the initial, biased impression. Another term for this phenomenon is *reconstructing credentials*, because we changed the perceived value and interpretation of each credential based on whether the person matched expectations or not. This process is especially worrisome because, in the end, we've come up with an explanation for the final decision that *sounds* well thought out, rational, and reasonable. The decision *seems* to be based on objective criteria, not biases or opinions or stereotypes. But the whole decision process is compromised by the initial, biased impression. **Thus, impression justification inadvertently supports discriminatory outcomes.**

Remember that each of these bias habits is *generalizable* across different target groups and *customizable* to different contexts. Impression justification isn't *just* about gender bias in police chief hiring decisions! Recall the Hurricane Katrina news captions I described in chapter 1, in which the Black boy was characterized as "looting," but the White couple was characterized as "finding" supplies. Think about how impression justification could have led to those captions going to print: Maybe the caption writer gave the editor a few captions to choose from, and impression justification led to the editor choosing to run the picture of the Black boy paired with the caption saying he "looted" supplies. The stereotypic expectation is to think of Black boys' behavior as criminal, so when the editor sees the criminal wording with the Black boy's picture, that pairing matches the expectation, which gives the editor a "good gut feeling"—that caption seems to make sense and "fits" the picture well. As such, it's more likely that the editor likes and chooses that caption, leading to it going to print.

The gut feeling that initiates impression justification can also arise from other sources besides stereotypic expectations. For instance, if you find someone physically attractive, or you find out you have something in common with them, that could also give you a good gut feeling that ends up coloring your subsequent judgments. **Start tuning in to the various ways impression justification can occur!** It's for you to look around in your life and your work to see some of the places this bias habit (and the others we'll learn about) might play out.

NORM ENFORCEMENT

In addition to setting up expectations about what people *are* like, stereotypes also become norms, meaning they set up expectations about what people *should* be like. Stereotypes tend to say that women are loving, nurturing, and passive, which becomes a norm saying that women *should* be loving, nurturing, and passive. Stereotyped norms about men say they should be physically strong, and they should not be prone to being emotionally sensitive. In other words, stereotypes and norms are both *descriptive* (saying how things allegedly *are*) and *prescriptive* (saying how things allegedly *should* be). These norms are often unspoken, but we behave in ways that enforce them, which is why this bias habit is called **Norm Enforcement**.

A crucial piece of norm enforcement lies in what we learned earlier about how we experience frustration and irritation when our habits get disrupted. **When people violate a norm, we feel irritated toward them.** We just don't like them as much, and we can do things to penalize them for violating the norm. Think about a boy in third grade who isn't very physically strong and is prone to being emotionally sensitive. He's violating the stereotypical norms about boys and men, which will irritate some of his classmates, who might call him names, saying he's a wuss or sissy or "not a real man." Those insults are penalties for the norm violation, which, intentionally or unintentionally, push him to conform with the normative ideals of manhood.

Similarly, when a woman is not focused on being loving and nurturing and is instead focused on her career and asserting her ideas at work, people often don't like her as much and may sabotage her behind her back. In fact, in recognition of the pervasiveness of this issue for women in the workplace, there's a whole research literature and abundant cultural discourse about how women can assert themselves and their ideas at work while still managing to be "likable." But this conflict does not occur the same way for men, because being assertive matches norms of how men are "supposed to" act, so we don't dislike assertive men the same way.

Norm enforcement, like most bias habits, can have a wide range of intensity levels. The initial emotional reaction to a norm violation could be as minor as thinking "That's weird" up to "Something's off about this person" to "They're so annoying!" all the way up to "I hate them!"

We also tend to like it when people conform to norms, behaving the way we habitually think they "should" behave, and we end up encouraging or praising that conformity. A pervasive way we see this play out in workplaces is in the types of tasks people get praised for doing. For instance, women and people of color end up doing "support tasks" more often than their male or White counterparts. When a work group needs someone to take notes for everyone at a meeting, or make coffee, women are asked to do those tasks more often than men who have the same title or level in the workplace hierarchy. There are, of course, historical reasons for this norm as well. Most workplaces were historically led by White men (and many still are), and when women and people of color began entering those workplaces in higher numbers, they more often filled support roles. As people became accustomed to seeing women and people of color working in support roles, that became the norm.

Another example comes from Dr. Nancy Grace Roman, a scientist who made many important contributions to the US space program. Dr. Roman shares this memory: "I still remember asking my high school guidance teacher for permission to take a second year of algebra instead of a fifth year of Latin. She looked down her nose at me and sneered, 'What lady would take mathematics instead of Latin?'" The norms, especially at that time (the 1930s), said that young women should be more interested in the humanities, not math and science. You can see in Dr. Roman's story how her guidance counselor felt annoyed at her for violating the norms, then made a move to enforce the norm by pushing her away from the math class.

One of my colleagues is a Black scientist and shares a similar story related to his race. He knew he wanted to be a scientist in middle school, but when he expressed an interest in pursuing a science career, many of his teachers would respond along the lines of "What? Don't be silly—you should go out for the basketball team!" Again, stereotypes about young Black men create a norm that says they should be more interested in basketball than science, and his teachers did and said things to try to enforce that norm on him.

Luckily for human knowledge, both of the scientists in these stories pushed back and overcame these normative pressures, going on to pursue impactful careers in their respective sciences. But I have to wonder: How many other great contributions have we missed out on because people were pushed away from a career in which they might have excelled?

Norm enforcement need not *only* arise from stereotypes that become norms. Any sort of habitual way of doing or thinking about things becomes the way we think things "normally" should be. Then disruption of those often unspoken norms will lead to frustrated emotions that may lead to lashing out at the cause of the disruption. Recall that we discussed some examples of this in chapter 2, such as when we need to disrupt our habitual talking or walking speed for someone with a disability, or when someone's default way of thinking about gender or sexual orientation is challenged by LGBTQ+ people existing and coming out.

An attendee at one of my trainings, Rebecca, shared a way norm enforcement played out in her life. Rebecca really enjoyed watching sports, especially the postgame in-depth analysis shows, in which sportscasters and coaches discussed a game minute by minute, analyzing all the details of how the game went and how it could have gone differently. She'd been noticing something in her mind and behavior for a while that bothered her. Whenever there were *women* sportscasters on these shows, she found that she just didn't like them. She'd catch herself thinking, and sometimes even saying, things like, "I wish she'd just shut up and let the men talk!" It wasn't that Rebecca disagreed with the points the women were making; she just felt hostility toward them. She knew this was "some kind of" bias or sexism, and it bothered her. After all, Rebecca herself was a woman who loved hearing about and discussing sports. So why did she hate these professional women sportscasters so much?

The answer came when she learned about norm enforcement, and it "clicked" for her. The stereotypes say that men talk about sports, not women, and that was operating as a norm for her. When these women sportscasters violated the unspoken norm by doing their jobs, Rebecca felt frustrated and irritated toward them as a result. This realization was very powerful for Rebecca, because it meant that rather than just having a vague idea that there's "some kind of bias" happening, she now understood how and why these feelings were arising in her. She decided that from that day on, when she felt those irritated emotions, she would consciously redirect them. Instead of being annoyed at the women sportscasters, Rebecca was going to be annoyed at the *norm* and start saying things like, "It pisses me off that people think women can't talk about sports! That's such a dumb norm!" In this way, Rebecca **put her frustrated feelings to work *against* the bias habit**, instead of perpetuating it.

Recall what I said in this unit's introduction, about how, like Rebecca, we often experience "bias" as some vague sense that something is wrong. In Rebecca's case, having a step-by-step understanding of this particular bias habit, and a label for it, empowered her to disrupt it. That's what we want to start doing with all these bias habits!

Rebecca's story provides an opportunity to emphasize another key point in our journey to break bias habits. You'll recall that I said at the outset that the skills we're learning are **generalizable** to various potential targets of bias and **customizable** to your various unique contexts. I don't watch sports recap shows very often, so none of my personal examples would ever deal with bias against women sportscasters in the same way as Rebecca's example. But in applying the skills she learned during this training, Rebecca generalized norm enforcement to see how it applied to bias directed at women sportscasters, customized into her specific life context of watching and engaging with sports shows.

I want you to do the same for your life contexts! I'm not the expert in your life, your work, your play. ***You* are the best expert in knowing your own life**, and it's for you to look around and apply the skills we're learning to identify where these bias habits might play out.

SELF-FULFILLING PROPHECY

Our final bias habit in this chapter is something you're most likely familiar with as a general idea in culture: **Self-Fulfilling Prophecy**. When we have expectations about how other people are going to behave, we often act in ways that bring out the behavior we were expecting, getting people to live up to our heightened expectations or live down to our lowered expectations. Our stereotyped expectation becomes reality, through our own behavior.

There were sprinkles of this bias habit in our examples of the previous two habits. With impression justification, we discussed our good gut feelings about a man in the police chief role leading us to hire him over a woman, and our expectation becomes reality—the prophecy is fulfilled! With norm enforcement, we expect women to be in support roles, so we ask a woman to make coffee because it feels more "right" than asking a man—the expectation becomes reality through our own behavior. But self-fulfilling prophecy *also* deserves acknowledgment as a bias habit on its own; our expectations can also "become true" through our behaviors without those emotional components.

Self-fulfilling prophecy has been studied in a plethora of ways in decades and decades of research! If I tell my lab assistant running one of my research experiments that I hypothesize that people in Group A will smile a lot, but people in Group B won't, just planting that idea in my assistant's head will likely lead them to smile more at Group A, which will make the people in Group A smile back. My lab assistant behaves in the way they think is appropriate for the group, and that makes the expectation become true!

Studies in the 1960s showed this phenomenon with students training rats to run mazes. Half the students were told their rats were bred to be especially smart at mazes, and the other half were told their rats were bred to be very bad at mazes. In reality, all the rats came from the same breeding stock. When students believed their rats would be smart, they played with them more, encouraged them more, and petted them more, and the rats actually did become better at running the mazes. The students who believed their rats were dumb didn't give them as much love and attention, and their rats ended up being worse at the mazes.

An attendee at one of my training sessions, Qisti, shared an example of self-fulfilling prophecy from her life. Many years prior, when she was a graduate student, she was in a class that required her to learn how to administer and score IQ tests for children. Several of her close friends let her practice the IQ tests on their kids. When she scored those tests, it turned out that her friends' kids were exceptionally gifted. It was somewhat of a surprise, but when she thought more about it, it made sense. After all, Qisti was a smart person, so her friends were smart, and of course that would mean their kids would be smart, too! So, she excitedly shared the news with her friends—their kids were gifted! After sharing that news, however, Qisti learned that she scored the tests wrong, accidentally bumping up the kids' IQ scores much higher than they actually were. She was so embarrassed by her mistake that she could never bring herself to tell her friends she had misled them. In the following years, because her friends believed their kids were gifted, they kept pushing their kids to get better grades, to take honors classes, and to go to top colleges. The kids ended up being very smart and very successful, spurred on, at least in part, by the incorrect IQ scores Qisti gave their parents.

When I shared Qisti's story at a subsequent training session, someone in the audience asked, "So, should we all just tell our kids they're gifted and push them like those parents did?" Before I even had a chance to respond, another audience member chimed in, applying ideas we'd covered, saying, "Well, no, because if your kid doesn't live up to the expectation, that disruption will make you feel frustrated and irritated toward them, like we learned earlier." I couldn't agree more! This response illustrates why the bias habit-breaking training is a collection of science-based *skills*, not just a list of scientific facts. We're talking about interacting and dynamic processes that we have to consider and figure out and customize to individual situations. **We must apply thoughtfulness and nuance to everything we're learning here.**

Self-Fulfilling Prophecy and People with Disabilities

A notable way this bias habit comes into play is with people with disabilities. Often in cases of people with disabilities, people create self-fulfilling prophecies because they're trying to be helpful, but doing something for someone else might interfere with their chance to show what they are capable of. It often comes from a genuinely kind, caring place, but the help isn't always wanted or needed—and may actually be detrimental.

Suppose you have a son who is injured at soccer practice and he'll have to use a wheelchair to get around for an extended period of time. Since the wheelchair is new to him, you expect that he won't be good at maneuvering around himself, so you push the wheelchair for him. By doing that, you've made it so he has fewer opportunities to develop the skills and muscles needed to move the wheelchair himself. You acted on your expectation, which created the outcome you expected!

I have a dear friend, Nicki Vander Meulen, JD, who is a lawyer and disability advocate. She's taught me a few guidelines in this area that I'd like to share with you. The first is to *always ask* before touching anyone's wheelchair, cane, or any other kind of assistive device. That may seem obvious to you, because most of us are taught as kids not to touch other people's things, but this issue is big in the disability community. *Many* people, trying to be helpful, will push someone's wheelchair without asking or retrieve someone's cane or walker from across the room for them without asking. For all you know, maybe the person *wants* to walk across the room without their cane, because they're building up their ability to walk without it. Maybe that action is part of their physical therapy. But if you do something for them without asking, you've taken away some of their autonomy, and perhaps their opportunity to build up their muscles or stretch their tendons, or whatever else they may have going on. The kindest thing to do is to *offer* help and give them the chance to accept or decline it.

These principles apply to developmental disabilities as well. In workplaces that employ someone with a disability, we see self-fulfilling prophecy play out a number of ways, but one of the most common occurs when work assignments are given. Suppose Claudio is handing out assignments to his employees. He tends to give Rhonda less challenging tasks because she has an anxiety disorder. Claudio's behavior robs her of the chance to learn, to grow, to rise to the challenge of harder tasks. She also then has fewer impressive completed projects to showcase her contributions at work. Claudio's expectations about Rhonda led him to behave in a way that made his expectations a self-fulfilling prophecy.

A handy phrase that my friend Nicki taught me for situations like these is to always start from a place where you "presume competence." For whatever position or job title a person holds, **presume they can competently do all the tasks normally associated with that job**, until they tell you otherwise. The person will know their own limitations and needs far better than you will as an outsider to their actual situation. They may need an accommodation to do certain tasks or may need to do them in a different way than other employees, or they may need more time. Alternately, they may not need anything different from the other employees! Listen to what their needs and wants are and work with them to make any needed accommodations. But start from a place of presumed competence. Anything less than that is setting them up to fall short and live down to the lowered expectations you're putting on them. **If you presume less than full competence, it can become a self-fulfilling prophecy.**

Onboarding and "Elite" Status

Another past attendee of the bias habit-breaking training identified a way self-fulfilling prophecy was playing out in his investment firm. Whenever the firm had a new hire who came from an "elite" business school, such as Harvard Business School, that person would receive special treatment. The hiring manager would introduce the new person to all the senior partners at the firm, say things like, "Hey, look at us! We got this Harvard alum! Aren't we fancy?," and tell the senior partners all the wonderful accomplishments and talents the new hire had. But when a new hire came from a more average business school, they would not get the same treatment.

It was readily apparent to the discussion group how this treatment would create an advantage for new hires from "fancier" schools. If the new hire from the "elite" school gets to meet and have face time with all the senior partners, they'll feel more welcome and connected at work than new hires from a more average school who don't get that treatment. Those connections, as well as hearing that people think they are a "superstar," could lead to them actually performing better. As the group discussed this, they further identified that, really, the firm should treat all its new hires as "superstars." After all, even if the new hire came from an average business school, that person was hired because the firm considered them to be impressive and thought they would succeed. We discussed this with the full audience, which included the firm's relevant policymakers, and as a result of this discussion, they immediately set a new policy. From that day forward, every new hire would go through a formal onboarding procedure, which would involve the hire being formally introduced to all the senior partners, with the hiring manager talking about all the great accomplishments and skills that led to the person being hired. The new policy called for every new hire to be treated like a "superstar" when they started at the firm.

This investment firm example has become my favorite story for illustrating self-fulfilling prophecy. In addition to reemphasizing the customizability and generalizability of these skills, it has served as a catalyst for many useful discussions and considerations in other training sessions. That story shows a very practical application of what the audience is learning; it brings up bias based on elite university status (generalizability), which is a bias target group previously absent from the training; and it brings up a specific context where bias can play out—onboarding procedures—which had not previously been emphasized in the training (customizability).

In a session with an academic department, that story served as a jumping-off point for faculty to recognize that in their onboarding procedures for new tenure-track faculty, they tended to pair new hires with a mentor who came from a similar school. That tendency set up a pattern where Ivy Leaguers worked more closely with other Ivy Leaguers, public university grads worked more closely with other public university grads, and so on—creating cliques that sowed division in their department. Those faculty took the experience of recognizing this self-fulfilling prophecy back to their department and started making efforts to counteract it. In another group, there was an attendee who had felt marginalized in her academic department because she was one of only a few faculty who weren't from an Ivy League school; she used that story as a way to speak up about her experiences with her colleagues. Yet another group used that story to talk about their lack of formal onboarding procedures and how they could implement some to start off new employees on a more equitable foot. Additional examples abound.

I've taken us on this brief tangent as a way to encourage you to think through similar issues for your contexts. Could elite university bias be relevant in your workplace? What are your onboarding procedures like when a new employee joins your team? Each story and example I share is more than *just* a way for you to learn habit-breaking skills; they are each an opportunity to think about how bias habits may play out for you. They're ways to start doing the work of identifying the bias habits in your life.

CHAPTER 3 SKILLS SUMMARY

- The awareness skills from this chapter involve tuning in to these three bias habits and understanding what they look and feel like, so we can disrupt them:
    - In **Impression Justification**, stereotypes give us a good gut feeling when someone fits expectations and a bad gut feeling when someone doesn't fit expectations. Then those impressions color how we interpret evidence in a way that supports the gut feeling.
    - **Norm Enforcement** involves stereotypes being used as prescriptive norms about how people should and should not behave. We dislike when people violate those norms and like when people conform to them, leading us to act in ways to enforce the norms.
    - Through **Self-Fulfilling Prophecy**, our expectations lead us to behave in a way that causes others to behave the way we expected.
- These awareness skills are *generalizable* to various target groups and *customizable* to different situations and contexts.
- Learning to tune in to these bias habits and to recognize when they may be affecting our reactions to and behaviors toward someone is an *actionable* way to address them.
- The practice of paying attention to bias habits is also *self-sustainable,* because as we start noticing these bias habits, we'll get better and better at catching ourselves.
- We'll also be able to apply some of the tools we learn later to help in disrupting each of these three particular bias habits.

REFLECTION ITEMS

- You have just learned about three bias habits (Impression Justification, Norm Enforcement, Self-Fulfilling Prophecy) that can lead to disparities. Think of some occasions when each of them happened in your life. If you can't think of a real-life example, come up with a hypothetical situation you might encounter.
- Can you think of times you may have had, or could have, a good or bad gut feeling about someone based on whether they "fit" stereotypes?
- Have you ever felt a pang of annoyance, frustration, or mistrust when someone violated a norm?
- What are some ways self-fulfilling prophecy might play out for you?
- See if you can generate examples that illustrate how this chapter's content is
    - *actionable* (What are you *doing*?)
    - *self-sustainable* (How will you maintain it over time?)
    - *generalizable* (Consider examples related to race, gender, sexual orientation, age, disability, politics, religion, body size, or other groups.)
    - *customizable* (How might you use these skills in your various work, life, and play contexts?)

SHARE YOUR REFLECTIONS!

Your examples and experiences can contribute to our research, help us improve this training program, and provide guidance to others who complete this training in the future! Consider sharing your examples, insights, and ponderings with us at BiasHabit.com/share.

4

BIAS HABITS THAT PERPETUATE COGNITIVE INERTIA

AS HUMANS, we like to think of ourselves as very rational, objective beings. In an ideal but fictional world in which we were actually highly rational, when presented with new evidence that opposed our preexisting mental habits, the habits would change based on the new evidence. A cursory review of modern society makes it clear that this ideal is not our reality. People all too readily believe in conspiracy theories, superstitions, and verifiable lies that spread in culture. There are, of course, conscious and motivated mechanisms that contribute to this inertia (e.g., consciously choosing to follow a news outlet that shares your viewpoint), but there are also less intentional mechanisms (i.e., bias habits), which will be our focus in this chapter.

In chapter 1, we learned about how ideas have *cultural inertia*—once introduced into the zeitgeist, those ideas tend to persist. The same thing happens with ideas and habits in our minds—they have **cognitive inertia**. Cognitive inertia occurs because once something has been learned, especially something so well learned that it becomes habitual, it is more efficient for the brain to maintain that learned association than to change it, and our brains favor efficiency. It literally requires more energy to change established neural pathways, and our brains want to avoid expending that energy. This cognitive inertia is a large part of what makes habits hard to break.

In this chapter, we're going to learn about three bias habits—attentional spotlight, confirmation bias, and untested assumptions—that contribute to cognitive inertia. Over time, these bias habits cause the stereotypes to get stronger in our minds, making it more likely we will rely on them, even in the face of evidence that, rationally, shows the stereotypes to be untrue, inaccurate, or even just *less* true than the stereotypes say. These bias habits all have the net effect of making us feel like the stereotypes are more true—they perpetuate the stereotypes in our minds, regardless of the objective evidence.

ATTENTIONAL SPOTLIGHT

The first of these bias habits is called **Attentional Spotlight** and involves how our brains direct our attention. **Our attention gets drawn more strongly toward evidence that confirms stereotypes** and gets drawn away from evidence that disconfirms stereotypes—like a spotlight in a dark room, illuminating some items in the room and not others.

Remember what we learned in chapter 2 about how stereotype confirmation is rewarding. One fundamental feature of reward learning is that it influences our attentional processes—our attention is drawn toward things that are or will be rewarded. While writing this book, I witnessed a handy and relatable example of this process while at a dog-friendly restaurant with my friend and his dog, Rigby. The restaurant's owner, Rolando, had treats in his pocket, and every once in a while, he would give Rigby a treat when he walked by our table. Pretty soon

Rigby's eyes followed Rolando wherever he went. Rigby had learned there was a chance for treats whenever Rolando was nearby, and Rigby's attention became fixated (i.e., *biased*) in Rolando's direction.

Human brains are more complex than dog brains, but many of the same low-level learning and attention principles play out for us as well. In fact, many research studies on bias habits in attention use various methods of eye tracking, in which a computer program tracks where participants' eyes look to understand how stereotypes are guiding their attention, just like how I observed Rigby the dog's eyes following Rolando. People's eyes and attention are drawn toward confirmatory evidence and away from disconfirmatory evidence.

Attentional spotlight contributes to cognitive inertia by making us notice confirmatory evidence more often, regardless of the objective pattern of evidence out in the world. As we go through our lives, attentional spotlight will draw our attention such that we notice more and more confirmatory evidence and overlook many more cases of disconfirmatory evidence. Over time, therefore, it seems like the stereotype is true more often, because of the bias habit in our attentional processes. The stereotype is strengthened in our mind not because the stereotype is true in reality, but because the evidence we notice is biased by attentional spotlight.

One classic research study powerfully demonstrated the effect of attentional spotlight related to stereotypes of grade schoolers coming from low-income versus high-income families. The stereotypic expectation here is that people largely expect students from a rich background to be more advanced in school and students from a poorer background to be less advanced in school. Participants learned about a young girl named Hannah, who was going to be entering the fourth grade. Half the participants learned that Hannah came from a more affluent family, and the other half learned that Hannah came from a poorer family. Then the participants watched a video of Hannah taking a test. They all watched the same video, in which a teacher asked

Hannah questions and Hannah answered them. Hannah's performance on the test was mixed. There were difficult questions she got right and difficult questions she got wrong. There were also easy questions Hannah got right and easy questions Hannah got wrong.

After seeing this mixed performance test, the participants were asked to estimate Hannah's ability level. If the participants believed that Hannah was from a rich background, their attention got drawn more strongly to the evidence that supported the expectation that Hannah would be high performing. They pointed to the difficult questions Hannah got right as evidence that she was especially smart for a fourth grader. But they overlooked those easy questions that Hannah got wrong. Participants in the other condition, who believed Hannah was from a poorer background, had the opposite expectation, thinking that she would be lower performing. And their attentional spotlight drew them to the opposite conclusion; they were more likely to notice when Hannah got easy questions wrong and overlooked the difficult questions she got right. This pattern led participants who thought Hannah was poor to say that Hannah had much lower abilities overall. Even though these participants saw the *exact same* test performance, their attentional spotlights led them to selectively notice evidence consistent with their expectations based on whether she was rich or poor. The bias habit operated on how their attention was drawn more strongly to evidence that confirmed their expectations and away from evidence that disconfirmed their expectations.

Another research study was conducted with lawyers. The researchers asked the lawyers to review a law brief that had several errors in it. As you can probably surmise, errors can be very consequential in legal documents. The lawyers were led to believe the brief was written by either a Black man or a White man. They spotted more of the errors when they believed it was written by a Black man but didn't catch as many when they believed it was written by a White man. The stereotypic expectation here is that Black men are less competent and smart and will therefore make more errors than White men. Attentional spotlight, therefore, led them to notice more of the brief's errors when they thought a Black man wrote it.

This example highlights how bias habits operate to create disadvantage or to create advantage. In this study, the ideal, "objective" outcome would be to find all the errors. Therefore, the pattern we see when the brief was ostensibly written by a Black man is technically closer to the ideal scenario in which we catch all the errors. In this case, attentional spotlight created an advantage for the White man, in which people overlooked his mistakes. But this advantage for the White man's reputation is a disadvantage for the law firm overall, in that they may miss mistakes with potentially serious consequences.

A common way attentional spotlight plays out in workplaces, rather than just studies about workplaces, involves whose ideas get attention. In fact, this example happens so often that you've probably seen it portrayed in sitcoms: A group of people will be working on finding solutions to a problem, but when a woman in the group speaks up with an idea, people overlook her. A few moments later, one of the men speaks up with essentially the same idea, and then everyone notices what a great solution "his" idea is. Again, this is attentional spotlight. We unintentionally tend to expect good ideas to come more from men than women, so our attention is drawn more strongly to a proposed solution when a man speaks it. In response to this example, you may think, "Well, couldn't it also be that we notice the idea more the second time we hear it?" My answer to that is yes—absolutely! That *also* contributes to this outcome. Repetition fuels recognition. With all these bias habits, I'm never saying that bias is the *only* process that contributes to the outcome we observe. **Basically nothing in human behavior is driven *only* by one underlying process.** Many things interact to produce the outcomes we observe. What we're learning about here is how to **tune in to the ways bias habits are one of those many factors**.

CONFIRMATION BIAS

Whereas attentional spotlight relates to what evidence we notice, **Confirmation Bias** involves what happens after our attention has been drawn to evidence. Culturally, *confirmation bias* has a broad definition that may connote different things; for our purposes, I'm going to define it as **an imbalance in how we learn from confirmatory versus disconfirmatory evidence**, after our attention has been drawn to it.

Recall again that expectation confirmation is rewarding, and expectation disconfirmation is aversive. There's a fundamental imbalance in the strength of learning from reward versus aversion; under typical circumstances, reward has a much more powerful influence on our learning than aversion. If you've ever heard that kids learn better from praise than punishment (i.e., favoring the carrot over the stick), it's that same idea. What this means for us in this chapter is that **each piece of confirmatory evidence holds more weight** in our learning and memory than each piece of disconfirmatory evidence. In other words, when we encounter people who happen to match a stereotype, like Black men who are tall, they exert more influence in our minds than people who violate that stereotype, like Black men who are short.

In a series of confirmation bias and stereotyping research studies in my lab, the imbalance between confirmatory and disconfirmatory evidence came out to a rough numerical ratio of 3:1 favoring confirmation. If we think of this pattern in terms of how much figurative weight we give each piece of evidence, it means that each time we encounter a person who confirms the stereotype, our minds give them at least three times as much weight as each person who disconfirms that stereotype. In other words, it takes at least three people who disconfirm a stereotype to balance out the influence of one person who confirms that stereotype.

To be clear, this 3:1 ratio comes out of how we designed that series of studies; I don't want to mislead you into thinking that this ratio is some

CONNECTIONS: CONFIRMATION BIAS AND SELF-FULFILLING PROPHECY

Think about how confirmation bias can operate in conjunction with self-fulfilling prophecy. For example, an attendee at one of my trainings, Yoichi, told me about how people often assume that he'll be very serious and emotionally cold because he's Japanese. When he met new coworkers, they would behave in very cold and formal ways toward him, which Yoichi then returned in kind—their assumptions about him thereby became self-fulfilling prophecies. Then, his more emotionally cold behavior became confirmatory evidence. If one of his coworkers ever questioned their assumption about Japanese people being cold, their experiences with Yoichi as a confirmatory example would exert more power in their assessment than any other Japanese people they had encountered who were more emotionally warm. Cases like these often make it so that we feel, think, or believe stereotypes are more "objectively true" than they are, sometimes even drawing on confirmatory evidence that our own behavior created!

sort of magic number or universal truth. The exact "weight" of the imbalance will vary for different situations and stereotypes. Many people, however, find the 3:1 ratio a handy rule of thumb in practice—when exposed to one piece of stereotype-confirming evidence, they try to think of, or expose themselves to, three or more pieces of disconfirmatory evidence to try to override confirmation bias. I also think that is a handy rule of thumb, and we'll revisit similar ideas later in unit 3.

For now, the principle I want you to learn is that there *is* an imbalance, such that **our brains tend to give more weight to confirmatory evidence than disconfirmatory evidence**. Even if we overcome attentional spotlight and force ourselves to notice all the evidence we encounter, confirmation bias will give extra weight to the confirmatory evidence in memory, so the stereotype will grow stronger in our minds over time.

Confirmation bias and attentional spotlight operate in concert to perpetuate stereotypes, making them stronger in our minds and more likely to influence us in ways we don't want them to. **Attentional spotlight involves what evidence we notice**, whereas **confirmation bias involves how much different types of evidence influence us** once they have been brought to our attention. Our next, and last, bias habit under the banner of cognitive inertia involves what happens when we have no evidence at all.

UNTESTED ASSUMPTIONS

We've talked a lot about when a stereotypic expectation is confirmed or disconfirmed, but there's another possible outcome. People can have expectations that get neither confirmed *nor* disconfirmed; these are **Untested Assumptions**. In fact, in everyday life, this outcome is probably the most common. An expectation pops to mind, but we never learn if it was correct or incorrect.

One of the most common types of racial stereotyping we see on college campuses is people assuming that Black students are on sports teams and that they may be attending college on sports scholarships. Black students report that others make these types of assumptions about them all the time, and many students (of various racial groups) admit that these assumptions often pop into their minds *about* Black students.

Suppose that Erika is a college student, and on the first day of class, she notices that one of her classmates is a Black man. Upon seeing him, she thinks to herself, "Oh, I bet he's on the basketball team!" What happens next? Do you think she flags him down after class to find out if her assumption was correct? Probably not. She probably just goes to her next class. Officially, her assumption remains untested. If she remembers it later on, however, how does she think about it? Does she think, "Oh, I saw my Black classmate and assumed he was on the basketball team, but I never learned if I was right or wrong, so I don't really know"? Again, probably not. More likely, she'll think, "Oh yeah, I saw that basketball player earlier," **treating the assumption like it was a fact**.

This sequence of thoughts is very common, perhaps the most common way stereotypic thoughts happen. We encounter someone, and a stereotypic assumption pops to mind about them, but it isn't necessarily *important* to us in the moment, so it goes by without much thought on our part and is never challenged or tested. Little untested assumptions pass through our minds all the time. Although they aren't confirmed, we tend to passively treat these assumptions as if they're confirmed. You can think of this like a form of "soft confirmation"; in reality, it's neither confirmation nor disconfirmation, but without the disconfirmation, we generally think the assumption is "probably true." Over time, these untested assumptions contribute to cognitive inertia because **they strengthen the stereotypes in our minds**.

How do untested assumptions strengthen stereotypes in our mind? It's a form of mental rehearsal. In grade school, did you ever memorize multiplication tables by rehearsing them over and over in your mind? Each time we make a quick assumption, that's like a rehearsal of the idea, strengthening the stereotype in our minds. I sometimes do a demonstration of this with students or audience members: I ask them to recite "two times two equals five" aloud several times. After that recital, I continue talking for a few moments, then I suddenly point at a random audience member and surprise them by abruptly asking, "Quick! What's two times two?" and, startled, they answer "Five!" I joke with them a little, explaining that just because we *rehearsed* saying that two times two is five, I didn't mean for them to *believe* it. Even when we very much *know* something is untrue, mere mental rehearsal makes that idea stronger in our minds.

Put another way, untested assumptions exist in the "maybe" space, between confirmation and disconfirmation. But those "maybes" tend to lean toward the side of "maybe yes" rather than "maybe not." In terms of what gets "programmed" into our cognitive systems, our brains don't really pay good attention to qualifiers (e.g., "maybe," "possibly," "perhaps") added to statements. The statement itself just gets written into our minds. Going back to our basketball player example, even if Erika was very careful to think, "*Maybe* he's on the basketball team," it still gets treated in her brain basically like a confirmatory statement, strengthening the association that assumes Black students are basketball players. This "soft" confirmation isn't as strong as if she got true, actual confirmation, but it is still a rehearsal that strengthens the stereotype in her mind.

In many cases, we also *act* on our untested assumptions as if they were true. Imagine that Jake is inviting some of his coworkers to a Super Bowl party. He invites the people he thinks would be most interested in football. When he considers inviting his coworker Kevin, the stereotypic assumption that comes to mind is that because Kevin is gay, he probably doesn't like sports. So Jake doesn't invite him. The next time he has a party to watch a sporting event, he thinks about "who likes sports" based on who came to his Super Bowl party. Jake's ideas about who does or doesn't like sports get reinforced by how he acted on his own assumption.

In another situation, suppose Rita sees a Black man walking down the street toward her, and she has the quick assumption that he might be dangerous, then crosses the street to avoid him. In this case, she acted on her assumption, and that action actually prevented her assumption from potentially being tested. Maybe if she stayed on the same side of the street, she would have gotten closer and overheard him on the phone giving kind, loving advice to a friend, seen a badge identifying him as a nurse, or something else about him would have suggested that he wasn't dangerous. As it stands, the assumption is untested, but it has now been rehearsed in her mind and through her behavior, reinforcing the idea of Black men as dangerous, to continue influencing her in future encounters.

Untested assumptions come up particularly often related to stereotypes about the LGBTQ+ community. People often use LGBTQ+

> **CONNECTIONS: UNTESTED ASSUMPTIONS AND SELF-FULFILLING PROPHECY**
>
> Consider how untested assumptions are synergistic with self-fulfilling prophecy as well. Maybe Norma assumes that her employee, Dana, won't be very good with technology because Dana is an older woman. Because of her untested assumption about Dana's alleged lack of tech ability, Norma doesn't assign Dana tasks that involve computers, which means Dana never gets a chance to develop or showcase those skills.

stereotypes as categorization cues. For instance, people think gay men are fashionable, so if they see a man who is dressed well, they assume he's gay. Similarly, people will assume women with short haircuts are lesbians or that women with facial features more associated with masculinity are transgender. And so on. Because the defining features of LGBTQ+ group memberships are not generally visible (i.e., you can't see whether a man is attracted to other men or not), people use stereotyping to make assumptions about those identities. You may have heard this type of stereotyping called *gaydar*, which is a portmanteau of *gay* and *radar*. How might this kind of stereotyping play out?

Imagine a group of friends sitting at an outdoor café, watching passersby. They see a fashionably dressed man and think (or say), "I bet he's gay!" But they don't run up and ask him; he walks by, and they presume they were correct. This sequence happens again and again, while they're sitting at the cafe, or over several weeks or months as someone is just going about their life. A man's appearance prompts a quick assumption about his sexual orientation, but the assumption is rarely or never tested. Over time, they really start believing they can "tell" who is gay based on looks, but in truth, there was *never* any actual evidence, just untested assumptions!

As a gay man myself, I'd prefer people just ask me if they want to know about my identity—or better yet, let me reveal it myself if and when I feel comfortable doing so! Nevertheless, people very frequently use appearance stereotypes to jump to conclusions about who is LGBTQ+. Overall, however, these judgments are very inaccurate. In fact, because LGBTQ+ people make up such a numerically small percentage of the total population, the *majority* of people who "seem gay" based on appearance are actually straight! There are more straight people who "seem gay" based on stereotypes related to their clothing, hair, or other aspects of their appearance than there are gay people *in total*. And not all gay people "seem gay." For many people, untested assumptions are a bias habit that fool us into thinking we "have good gaydar," when **in fact we mostly have no evidence at all**! We're just stereotyping.

An attendee at one of my training sessions shared an example of untested assumptions in her life. She was a medical doctor, and one day on her way home from work, she went grocery shopping. At the store, she saw a man who looked a little off. He was a bit underweight, had dark circles under his eyes, and his clothes were disheveled. Based on what she saw, her immediate judgment of him was that he was a needle drug user. To her, this assessment meant he might be dangerous, so she didn't want to get close to him, and she avoided going down any aisle she saw him in. After that night, whenever she went back to that grocery store, she would remember "the needle drug user" and hope that he wouldn't be at the store this time.

When this attendee learned about untested assumptions, she thought back to that man. Upon closer examination, she recalled that she didn't see any "track marks" on his arms, which would be a direct indicator of needle drug use. And, being a doctor, she could think of other things that might be going on with the man, based on what she *did* see. Maybe he was a cancer patient—the dark circles under his eyes and low body weight could have been the result of cancer treatments. Or maybe he was just poor, or anorexic, or . . . she was able to generate many other medical and lifestyle explanations for the handful of things she had been able to observe about him, none of which meant he was dangerous. But in the original encounter, it hadn't been her job to diagnose what was going on with him. A quick assumption was good enough to go about her shopping. But from that day on, she remembered her assumption like it was a fact, and it essentially rehearsed itself each time she went back to that grocery store and remembered "the needle drug user."

I have a similar story from my own life. I was sitting at a table in a bar with some friends, when I noticed a woman had started walking toward us. As she walked, she stumbled a bit. I thought, mildly annoyed, "Oh great, some drunk lady is coming to bother us!" I think assuming that someone stumbling in a bar is drunk would be a pretty typical thought for most people. It turned out that she knew my friends at the table and she joined us, then I got to know her a little bit. It also turned out that, in fact, she was not drunk—she has cerebral palsy, which means she's often a bit unsteady on her feet.

I'd like to say that I was so thoughtful and clever that I tested my own assumption and found it to be incorrect, but I'd be lying if I said that. I just happened to be lucky, and the circumstances happened to reveal that my assumption was incorrect. That woman ended up becoming one of my very best friends. But I want you to think about what would have happened if I hadn't met her. The next time I saw her around town, she *also* would have been a bit unsteady on her feet, and I probably would have thought, again, that she was drunk. If I saw her a third time, that thought would exacerbate itself, and I might start thinking, "Wow, this lady is *always* drunk." The untested assumption of her being drunk would have perpetuated itself in my mind, and I would have been making unfair judgments about a delightful person who would have otherwise become one of my best friends! I got lucky in this case, but how many times have you or I missed out because of an untested assumption?

CHAPTER 4 SKILLS SUMMARY

Each of this chapter's three bias habits contributes to cognitive inertia, strengthening stereotypes in our minds in a way that is disconnected from a rational accounting of evidence.

- **Attentional Spotlight** biases our attentional processes toward noticing confirmatory evidence more than disconfirmatory evidence.
- Once our attention has been drawn to evidence, **Confirmation Bias** leads our minds to weigh each piece of confirmatory evidence more heavily than each piece of disconfirmatory evidence in memory.
- When we make assumptions but get no evidence about their veracity, those **Untested Assumptions** are a type of mental rehearsal that strengthens the stereotypes in our minds.

UNIT 2 WRAP-UP: SYNERGY AMONG BIAS HABITS

The **six bias habits we learned about in this unit can be synergistic**, working in concert. For example, think about stereotypic expectations that say men are better at science than women. Suppose we're hiring for a scientist position. As we look through the applicants, attentional spotlight might draw our attention more to the men's applications and away from the women's. As we're figuring out who to hire, maybe we engage in some impression justification that leads us to hire a man over a woman. The inherent assumption about the unhired woman is that she wouldn't have been good at the science job. But since she didn't get hired, we never find out what her abilities actually are, meaning it's an untested assumption! But we assume we were probably right.

Sometime later, we hire two interns, one man and one woman. We expect the man to be good at tasks more central to this group's area of

TABLE I
Six Bias Habits to Disrupt

Bias habit	Definition
Impression justification	Stereotypes give us a good gut feeling when someone fits expectations, and a bad gut feeling when someone doesn't fit expectations. Those gut feelings become the lens through which we interpret evidence.
Norm enforcement	Stereotypes become norms about how people should and should not behave. Then we act in ways to enforce those norms.
Self-fulfilling prophecy	Expectations lead us to behave in a way that causes others to behave how we expected them to act.
Attentional spotlight	Stereotypes lead our attention to stereotype-consistent information and away from stereotype-inconsistent information.
Confirmation bias	We give confirmatory information more weight than disconfirmatory information.
Untested assumptions	Without evidence, our assumptions get treated as if they were confirmatory information.

science, so we assign more of those tasks to him, giving him many chances to develop his skills, making that expectation a self-fulfilling prophecy. The woman intern gets assigned more "support" tasks, and when she tries to reach beyond those to do more scientific tasks, it bugs us, because she's violating the norms in our minds, and we push her back to support tasks, engaging in norm enforcement. Over time, as we look around, we see that the good scientists in the work group are all men, and people doing support tasks are women! Those are pieces of confirmatory evidence, which through confirmation bias will make the stereotypes stronger. Thus the biases persist.

I obviously constructed this hypothetical to highlight all six bias habits we've learned together in one scenario. I'm not trying to say that these habits are the *only* thing driving our decisions, nor am I saying that every bias habit will happen every time it can happen, nor am I saying that the bias habit will account for 100 percent of each decision.

***Of course* other factors come into play**—in fact, that's a central point of this book! These bias habits are powerful, but there's a lot we can do to disrupt them. Nevertheless, the constructed scenario above is not implausible. Remember that **these bias habits play out bit by bit over time, compounding each other's effects**.

REFLECTION ITEMS

- Attentional Spotlight, Confirmation Bias, and Untested Assumptions are all ways that your brain perpetuates biases. How might these play out in your work or life contexts?
- Can you think of a time you overlooked or underweighted disconfirmatory evidence?
- Have you ever acted on an assumption without knowing whether it was true? How might you catch yourself and test assumptions like this in the future?
- I discussed how untested assumptions can lead people to think they have good "gaydar." How might attentional spotlight and confirmation bias also perpetuate that kind of stereotyping?
- See if you can generate examples that illustrate how this chapter's content is
    - *actionable* (What are you *doing*?)
    - *self-sustainable* (How will you maintain it over time?)
    - *generalizable* (Consider examples related to race, gender, sexual orientation, age, disability, politics, religion, body size, or other groups.)
    - *customizable* (How might you use these skills in your various work, life, and play contexts?)
- Generate some additional ways the bias habits from this unit might exacerbate each other's effects.

- To wrap up unit 2, make sure you have a firm grasp of each of the six bias habits, which will come up throughout unit 3.

SHARE YOUR REFLECTIONS!

Your examples and experiences can contribute to our research, help us improve this training program, and provide guidance to others who complete this training in the future! Consider sharing your examples, insights, and ponderings with us at BiasHabit.com/share.

UNIT THREE

TOOLS TO DISRUPT BIAS HABITS AND CULTIVATE DIVERSE JOY

EACH OF THE BIAS HABITS and awareness skills we learned about in unit 2 will come back at various points throughout this new unit as we turn our attention to tools. **Each tool will help us in our efforts to disrupt bias habits and cultivate diverse joy.** Some tools will involve things like mental exercises to help reduce the influence of stereotypes in our minds. Other tools will involve more interpersonal activities, such as getting to know other people, making meaningful connections, and that sort of thing. These tools work together as a tool kit: **There's no one tool that is applicable in every single situation.** We're going to learn about how and why each of these tools can be useful and then figure out ways of applying them and integrating them into your life. **The tools are also synergistic**—working on one of them can help you in your application of others, so learning and using the full set of tools has compounding beneficial effects.

Remember that our goal is to **use these tools thoughtfully and intentionally**. For each of these tools, one *could* interpret them in a more superficial way, but my goal is for us to always be thoughtful and wise in how we integrate them into our lives, in service of our earnest intentions to disrupt bias habits and cultivate diverse joy.

5

FAVOR MINDFULNESS OVER BLUNT, INEFFECTIVE TOOLS

MINDFULNESS HAS BECOME a common buzzword within culture, and it's also a scientifically validated tool that can help deal with unwanted habits. Our goal here is not a comprehensive, in-depth exploration of mindfulness as a standalone concept, but its elements are threads woven throughout our journey. Approaches that focus on mindfulness emphasize that thoughts and mental habits often arise due to factors outside our control—for instance, stereotypes we've learned from culture. **The goal of mindfulness is to become aware of these thoughts as they happen and to let them pass without negatively affecting our behavior.** Does this sound familiar? That's because it's what we worked on in the previous unit! Developing those awareness skills helps us recognize many of the ways bias habits can influence our thoughts, emotions, and behavior. By becoming more aware of those bias habits, we can recognize and reduce their influence.

Mindfulness offers a better way to address mental habits than approaches that tend to be more blunt—and, in the end, ineffective. In fact, there are a lot of things people do that, intuitively, seem like they should reduce bias, but research shows **these intuitive tactics very often backfire and lead to *more*, rather than *less*, bias**. So I want to warn you away from these blunt, ineffective approaches. As we learn why these don't work, we'll see how a more mindful approach will be better for our efforts to disrupt bias habits.

DON'T TRY TO BLUNTLY SUPPRESS STEREOTYPES

When some people learn about how stereotypes can lead to biases in their behavior, their go-to approach is to just "push stereotypes out" of their minds! After all, their logic goes, if stereotypes are bad things, just don't think about them! This approach is a form of something called *thought suppression*, which tends to be ineffective across a number of domains.

A classic demonstration of how thought suppression backfires is to tell someone, "Don't think about an elephant!" Almost inevitably, the person immediately thinks of an elephant. Our brains don't process negations, such as "not" or "don't," very readily. The idea of an elephant first gets activated, and then we have to go back and try to "not" think of it. When we bluntly try to suppress thoughts in this way, it actually makes them *more* likely to influence us. They become more easily accessible and hyperactivated in our minds. In other words, they tend to "rebound." Even if we can suppress a thought for a short period of time, once we've stopped actively suppressing it, that thought or idea rebounds and influences our behavior more than if we'd never tried to bluntly suppress it. These processes occur for thoughts in general and for stereotypes and biases specifically. Just like other kinds of thoughts, **stereotypes rebound after they're bluntly suppressed**. Over time, bluntly suppressing stereotypes leads people to express *more* bias, not less.

If there's an elephant in the room, you should be aware of it. Maybe you don't want the elephant of stereotypes to drive your decision-making, but you at least need to know it's there so you can walk around it. **We're better off noticing when and how stereotypes pop to mind, then taking steps to reduce the likelihood that those stereotypes influence our actual judgments and behaviors.** In fact, so much of what we learned in the last unit was helping us become *more* aware of how these thoughts come to mind and influence our behavior. We want to notice the stereotypes *more*, not less.

DON'T TRY TO BLUNTLY IGNORE GROUP STATUSES

Sometimes you hear people say things like, "I just don't see race!" These people are attempting to bluntly ignore group statuses. There are a number of reasons this type of approach is ineffective. The first is that it isn't really possible in most common circumstances. In the case of visible cues to group membership, like skin tone, unless you're blind or have another visual impairment, you will see the color of someone's skin, which will activate ideas about race in your brain, including stereotypes and biases. Another issue with trying to ignore group statuses is that many people value their group identities as important parts of themselves. So saying, "I don't see race" to someone can be disrespectful if they are proud of that part of their identity. You're also precluding any consideration of how someone's identity might inform their life experiences and what they bring to the table. Our social group identities are one important part of what gives us all different, uniquely valuable perspectives.

Further, in many cases, ignoring a group status can be a pervasive form of bias against that group. For instance, LGBTQ+ people report that they commonly face exclusion because of pervasive assumptions that "everyone" is heterosexual and cisgender. In other words, they consistently encounter individual people or cultural ideas that ignore the existence of different sexual orientations or gender identities. Actively trying to ignore group statuses in the case of sexual orientation, therefore, doesn't actually ignore *all* sexual orientations; it perpetuates the notion that everyone has a heterosexual/straight sexual orientation. Imagine you're a woman starting a new job. You're being introduced to new coworkers all day, and they keep asking if you have a husband, when you actually have a wife. Wouldn't that get old fast? It would! Similar patterns play out for other social group identities as well—typically, **ignoring group statuses actually perpetuates assumptions about the dominant or majority group status as the default**, more "normal" identity.

Some people who try to bluntly ignore group statuses are genuinely trying to express that they don't want group statuses to unfairly *influence* their behavior. They think that if they ignore race, then race bias won't occur. But as we've already said, that's not really how brains operate. As with each of these blunt approaches, ignoring group statuses usually ends up being ineffective and producing more bias, rather than less. Instead of ignoring group statuses, we should be mindful of them and the thoughts, feelings, or expectations they may give rise to.

In my experience, most people would like others to take their group statuses and identities into consideration, even if it's only a little bit. They don't want people to completely ignore their identities, and they don't want people to see them *only* or primarily as any one identity. As a gay man, I appreciate it when people ask whether someone else has a partner or spouse, rather than asking a man if he has a wife or girlfriend. That's giving sexual orientation a small bit of consideration. Or if someone uses a wheelchair, it would actually be a little rude if you *completely* ignored their accommodations needs, for instance, by not paying attention to whether they can roll their wheelchair on the path you're taking together. It is more inclusive to give group statuses a small, appropriate amount of consideration.

For now, the key lesson is to understand that the mental approach of just **trying to bluntly ignore group statuses is neither possible nor**

WHAT ABOUT "BLINDED" HIRING PRACTICES?

To be clear, we've been discussing the *mental* practice of trying to ignore group statuses. You might also be wondering about instances in which you can truly not know what someone's group status is. For instance, if you need to hire someone with a certain set of math skills, you could set up an online application where people complete math problems, but you never see them or learn their name, and the hiring decision would be based purely on the number of problems they get correct. If those math problems are *truly* the only important criterion, and someone's personality, life experiences, or other accomplishments play no role, then *maybe* that setup would work. In most real-world situations, however, there are other concerns and considerations, and some part of the decision process will involve meeting the person and therefore learning about their visually apparent identities, at least. Someone's actual interpersonal skills, or lack thereof, will probably be part of *any* hiring decision.

desirable, and more often than not it backfires. As with the other blunt and ineffective approaches in this chapter, a better approach is to acknowledge that someone's group statuses are among the many influences on how you see them and to be mindful of how they might give rise to bias habits.

DON'T TRY TO BLUNTLY RELY ON PERSONAL OBJECTIVITY

Sometimes people will say things like, "Well, if being biased is the bad thing, I'll just be objective! After all, being objective is the opposite of being biased, right?" But, as with all the misguided approaches in this chapter, bluntly asserting objectivity backfires.

Part of the issue here is that, as we've learned in past chapters, **our brain's inherent mechanisms for processing evidence don't lead to an objective accounting**. The pattern of input our brains receive doesn't reflect "objective" reality: Our input is biased because of biases in the media and our social environments, as we learned in chapter 1, and also attentional spotlights that skew what evidence we notice, as we learned in chapter 4. And once input has moved into our minds, we also don't objectively process that evidence, as we learned throughout chapters 3 and 4. **Subjectivity is an inherent part of being human.**

We're also not very good self-evaluators—which is to say, we're not very accurate judges of the extent to which our judgments are based in "objectivity" versus subjectivity and bias. The skills in this book are a big step in the direction of understanding and undermining bias habits, but **"objectivity," like "perfection," is an imaginary construct**.

When someone bluntly adopts "objectivity" as their tool to being less biased, it backfires in a few ways. One is that they tend to question themselves less. If they think they're an objective person, after all, then the thoughts that pop into their heads must be objective thoughts! They end up trusting those thoughts *more*—even those arising from stereotypes—because they imagine themselves as objective. People adopting this blunt approach tend to show *more* bias, rather than less bias.

A related point involves people who think of "objectivity" as part of their personality or identity, such as some people in professions that emphasize objectivity as an important trait, like judges, scientists, or lawyers. **If we uncritically think of ourselves as highly objective people, we end up questioning our automatic thoughts less** and being less mindful of some of our bias habits, which results in more bias in our judgments and behavior.

I highly recommend everyone recognize objectivity for what it is—an imaginary ideal construct. **Objectivity isn't something that can or ever will exist inside a person.** The practice of *striving* for objectivity can be important and admirable, but we should never believe it has been achieved. Also, in many cases, subjectivity itself has great value; one's unique perspectives and experiences can bring new insights, creativity, and transformative solutions to a problem. The imaginary nature of objectivity is why legal systems have checks and balances and why science constantly updates itself based on new evidence; even the collection of that evidence involves careful procedures defined by the scientific method, then evaluated through peer review processes.

Mindfulness helps us recognize that our thoughts are *not* objective reflections of reality. Rather than bluntly asserting our objectivity, **we're much better served by acknowledging that our mental processes and decisions are, in fact, subjective**. Cultivating mindfulness to understand the various ways bias habits may influence our subjective assessments will help us disrupt those bias habits before they result in discriminatory behavior.

FAVOR MINDFULNESS (AND EVIDENCE-BASED TOOLS)

Let me reiterate that the three blunt and ineffective tools described in this chapter tend to be very common and intuitive for people. Many people try out these tools because they believe they will be effective at reducing bias. Unfortunately, many people who develop nonscientific diversity trainings recommend these ineffective tools to others, because they seem "nice" based on intuition, although the scientific evidence says otherwise. In fact, someone at one of my trainings shared with me that he had recently gone through a human resources training that specifically told him that when he hired people, he should make sure not think about stereotypes, try to ignore things like race and gender, and be as objective as possible—exactly the three ineffective tools I had just warned him away from! Nonscientific advice that backfires is widespread, not just in the realm of bias, but in things like mental

TABLE 2
Favor Mindfulness over Blunt, Ineffective Tools

Don't bluntly . . .	Instead, mindfully . . .
Suppress stereotypes	Notice stereotypes and consciously avoid allowing them to affect behavior
Ignore group statuses	Be aware of group statuses and how they may create bias
Rely on personal objectivity	Recognize that thoughts are subjective

health advice, fad diets, or social media exercise trends. Just as (I hope) you want to make your physical and mental health care decisions based on advice from experts who use scientific evidence, our efforts to reduce bias should likewise be evidence-based.

We're much better served by being *mindful* and noticing when biases come to mind. Once we notice a biased thought or assessment, we can then put into practice some tools that scientific evidence tells us will be effective at helping to counteract that bias. Mindfulness, in terms of dispassionately noticing the thoughts and feelings that come to mind, will continue to be a thread woven throughout the remainder of this book.

CHAPTER 5 SKILLS SUMMARY

- In your efforts to disrupt biases, **Don't Try to Bluntly Suppress Stereotypes**, **Don't Try to Bluntly Ignore Group Statuses**, and **Don't Try to Bluntly Rely on Personal Objectivity**.
- The three blunt, ineffective tools each involve, in one way or another, running away from the reality that we are indeed vulnerable to various forms of bias, and each sets us up for failure.
- We're much better off if we acknowledge the following simple realities:
    1. Biased thoughts do come to mind.
    2. We can indeed observe identities that might activate bias habits.
    3. Humans are subjective.
- Please, **Favor Mindfulness over Blunt, Ineffective Tools**. When thoughts come to mind, we can evaluate them without attachment, assess how bias habits may influence our reactions, and put into place more effective tools to prevent and disrupt those bias habits from influencing our behavior.

REFLECTION ITEMS

- Think of times when you or someone else tried to use one of these ineffective tools. What are some of the ways they can backfire?
- Can you think of some times you've tried to suppress a thought and found yourself thinking about it even more?
- Have you ever tried to ignore something (e.g., race) but still found yourself influenced by it?

- Can you think of a time you or someone else asserted they were objective, when their decision or behavior was probably influenced by subjectivity and emotions?
- See if you can generate examples that illustrate how this chapter's content is
  - *actionable* (What are you *doing*?)
  - *self-sustainable* (How will you maintain it over time?)
  - *generalizable* (Consider examples related to race, gender, sexual orientation, age, disability, politics, religion, body size, or other groups.)
  - *customizable* (How might you use these skills in your various work, life, and play contexts?)

SHARE YOUR REFLECTIONS!

Your examples and experiences can contribute to our research, help us improve this training program, and provide guidance to others who complete this training in the future! Consider sharing your examples, insights, and ponderings with us at BiasHabit.com/share.

6

TOOLS TO RETRAIN YOUR MIND

IN CONTRAST TO the ineffective tools in the last chapter, the remaining chapters of this unit focus on *effective* tools. The four tools covered in this chapter are focused on helping you retrain your mind, working against bias habits to bring your mental habits more in line with your conscious values and intentions.

DETECT, REFLECT, AND REJECT BIAS

Our first tool involves starting to notice and release bias in the moment. When bias occurs, either from within your own mind or from out in the world, we want you to be able to recognize it, understand ways it might create problems, and let it go. Doing this involves three steps as we work to **Detect, Reflect, and Reject Bias**.

Step One: Detect

First, you have to *detect* the biased thought, reaction, or portrayal and label it as biased. You might detect a biased media portrayal, like a television show that portrays a person of Latin descent as a criminal, or detect someone around you doing or saying something biased. The bias may also be something that pops up in your own mind—a reaction relating to one of the many bias habits we learned about in the awareness unit. You might notice a stereotypic thought, a snap judgment about someone, or a gut feeling that arises from expectation confirmation or disruption. **By applying the awareness skills of the last unit, you will start detecting bias** more often when it comes to your mind or when it happens in the world around you.

Step Two: Reflect

Now that you've detected the bias, you want to *reflect* on it. The default state is for our quick, automatic cognitive processing to just barrel through our minds without reflection. After all, the function of mental habits is to provide a quick, easy answer we don't need to think about; we make a quick assumption about someone and move on. If we want to change those automatic, unintentional bias habits, we need to marshal our more deliberative, thoughtful cognitive processing. You might reflect on where the stereotypic thought or idea came from, for instance, trying to think about media or life experiences that may have taught you that particular stereotype. Also, what might some of its consequences be? The latter question could cover the consequences out in the world generally and for your own mind and behavior specifically.

The particulars of how you reflect will depend on the circumstance. But the key is to **engage that more thoughtful, deliberative processing**. In every case, however, I want to recommend that you end the "reflect" step by asking yourself whether this thought, idea, or image is

Figure 5. Three steps to mindfully release bias.

something you want in your mind: Is it something you want to believe is true? Is it something you want to keep as a part of who you are? Is it something you want to continue living in your mind? Your answer to these questions leads us to the final step.

Step Three: Reject

Lastly, if the stereotypic thought is something you don't want to keep, you then mentally *reject* it. **You let it go.** Perhaps you think, or even say, "That's not something I want in my mind." You're not beating yourself up about it nor bluntly trying to suppress it. **You gently and unemotionally decide it's not something you want to keep.** Like a small piece of trash in your hand, you decide to throw it away. I want to acknowledge you might have some emotional reactions—guilt, anxiety, frustration—and those are natural, especially if this bias is something you actually said or did that might have negatively affected someone else. But the goal here is not to dwell on those emotions beyond their mere utility to help you notice the bias and work on correcting it.

Using the Tool

Now, let's bring the three steps together. Maybe you see someone of Latin descent portrayed as a criminal on television. You can detect that portrayal, labeling it as perpetuating the stereotype of Latin folks as criminals. You then reflect on where it comes from and how it might affect you. Perhaps you consider how, as a stereotype-confirming example, confirmation bias will lead it to have a stronger influence than other representations of Latin people who aren't portrayed as criminals. Then, you decide to mentally reject it, because you don't want that idea in your mind.

Maybe you're at the grocery store and see a larger-bodied person and impulsively think, "Oh, they're probably buying junk food!" You can detect and label that as a rude assumption, reflect on how body size isn't always the result of eating habits, and even if it were, it is an unkind assumption to make, and reject the thought as something you don't want in your mind.

I was working with some college students in my research lab, and there was a computer task that needed to be assigned to someone. One of my students dismissed the idea of her doing that task by saying, "I can't do computer stuff, I'm a girl!" The other students continued dividing up assignments without really noting what she said, but I paused the conversation and gently pointed out that not only did her statement reflect gender bias, but we were a *bias research lab*. I then led the group in a short discussion about the stereotypical notion that "women can't do computer stuff." That stereotype becomes a quick, easy answer, and in this case, perhaps a bit of a self-fulfilling prophecy—she assumed she couldn't, so she didn't try, so she'd never learn! I asked her if she'd ever *tried* computer tasks like the ones we needed done, and she hadn't.

We also discussed how that statement could affect others: Maybe other women in our group had no particular opinion about their own potential tech-savviness, but hearing that statement, had it gone unquestioned, may have nudged them to think they probably wouldn't be suited to doing computer tasks either. And finally, we discussed whether that idea—that "girls can't do computer stuff"—really represented what we believed, what we wanted to persist in the world, or what we wanted our lab group to stand for.

In other words, I guided our group through Detect/Reflect/Reject as a collective. I *detected* that her statement communicated bias and helped others detect that as well, then led the group as we *reflected* on the origins and consequences of that bias, and as a group, we *rejected* that bias as something we disagree with and don't want to perpetuate. In the end, that student chose for herself to try the computer task. It involved a learning curve, but it was indeed something she was *able* to learn when she put her mind to it!

Learning to Detect, Reflect, and Reject Bias has utility both as a stand-alone tool and in conjunction with the other three tools in this chapter. It is one of the most important skills in all this training, because if we don't notice (detect), understand (reflect), and spurn (reject) bias in our own minds or the world around us, we won't be able to make meaningful change! The next three tools will help us with additional ways to retrain our minds.

REHEARSE REPLACEMENTS

Our next tool for retraining your mental processes is to **Rehearse Replacements**. The goal here is to practice rehearsing a new thought or idea as a replacement to contradict stereotypic thoughts, assumptions, or portrayals. For instance, if you find yourself thinking "girls aren't good at computers," you can practice replacing that thought with "many girls are great with computers!" This tool builds very readily on the previous one. Once you've detected, reflected, and rejected a biased idea, you can rehearse an idea that is more in line with how you want to be thinking. As noted earlier in several places, we learn biases and stereotypes from repetition—repetition in the media and our social environments teaches us the biases, then they are further perpetuated by our own repetitions, via our untested assumptions. Rather than leaving our minds to these passive rehearsal processes, this tool involves **actively employing rehearsal in service of reducing bias rather than perpetuating it**.

In some ways, this tool is really a very simple, straightforward idea. If you don't want to think one way, practice thinking another way! This tool shouldn't be difficult, but **we do need to be intentional and put it into practice**. In chapter 1, I told you about how I used this tool when I wanted to change my verbal habit of saying "you guys" to something more gender inclusive. I started practicing saying "you all" instead, and as I rehearsed that replacement over time, it became my new verbal habit.

Recall the story about Jennifer, who mistakenly assumed the doctor she was meeting was a man rather than a woman. Jennifer recognized that her mistake involved the preexisting mental habit that assumed doctors must be men. She could then rehearse replacements as a way to counteract that mental association by mentally rehearsing the idea that "women are doctors too" several times in her mind. This rehearsal builds or strengthens the mental association between women and the doctor role. If Jennifer is successful, the next time she is in a similar situation, the idea that the man is the doctor will still come to mind, because we can never fully erase those well-learned mental associations, but the idea of women as doctors will *also* come to mind. Both ideas will be activated because of the time spent rehearsing the replacement. In that instance, therefore, it won't *just* be the easy assumption that the doctor must be the man.

This tool can also be verbalized and shared with others to help in your practice. One self-described "gym bro" told me a way he and his workout partners used this tool. One of their habitual ways of motivating each other while lifting weights was to shout things like, "Don't be such a girl!" This practice changed when one of them pointed out that he didn't like perpetuating the idea that girls are weak, and instead they started saying, "Be tough like a girl!" This replacement was a bit humorous, for them and others observing them, because people don't often expect a "gym bro" to push himself to be "like a girl." At first, it kind of felt like a joke when they said it, but as they kept repeating that phrase, it eventually became their rallying cry.

I like this example because it very straightforwardly demonstrates someone recognizing that the default habit reinforces an unwanted, biased idea, and then finding a way to replace it with something that rehearses an idea they believed was better. I also like to imagine the positive downstream consequences of this particular replacement. Maybe someone else at the gym hears their chant and asks them why they're saying that phrase, giving them a chance to explain that they don't like how society tells girls they're weak, so they're trying to reinforce a different idea. If these gym buddies chant, "Be tough like a girl!" every time they work out, several days a week, it starts becoming a part of their identity, such that in other contexts, each of them is more likely to speak up if someone does or says something that implies girls are weak.

Intention

This "gym buddies" example also provides a chance for us to reemphasize the role of intention for these skills. In response to that example, a cynic might roll their eyes and sarcastically retort, "Oh, so just invent some new gym chants and it'll fix your bias?" My answer to this query is both yes and no.

In the "yes" column: The fact that **we learn from repetition** is the most axiomatic, noncontroversial idea in all of behavioral neuroscience. Repeating and rehearsing a phrase, idea, or image builds pathways in your brain that make the repeated association stronger.

In the "no" column: I'm not suggesting that you "just" recite some chants without ever thinking more deeply about the related issues. We're learning to use this tool, and all these tools, intentionally, *in service of* our goals to think and behave more fairly and with less bias. The rehearse replacements tool is not *just* about the repetition, but also your conscious personal choice about what you want to rehearse. Rather than leaving your mind to passively rehearse whatever associations are thrown into it from the world, **you're choosing to rehearse associations that you want in your mind**—the gym bros wanted to change what they said about girls, and Jennifer wanted to change how she thought about doctors.

Replacement Types

So far the replacements we've discussed have been in the form of simple phrases, but you can also rehearse replacements in the form of general ideas or values you believe in (e.g., "that's not how I want to think about Asian people," "I don't want to make assumptions like that"). They can also take the form of questions you want to ask. For instance, if you find yourself making an assumption about someone, rather than letting it pass and allowing your brain to treat it like confirmatory evidence, as we learned about with untested assumptions, you can practice asking yourself, "How do I know?" or "What's the evidence for that assumption?" as a way to start **training yourself to catch and question assumptions and where they come from**.

Replacements can also take the form of images or stories. Recall our earlier example about seeing a television show that portrays a Latin person as a criminal. To work against that idea in your mind, you can practice thinking about examples of Latin folks who contradict the image of Latin criminals. They could be real people you know, famous people, or even just hypothetical people you imagine. Maybe you imagine a Colombian-American woman who is an advertising executive, a loving mother, and an avid cycling enthusiast. The exercise of picturing her, and what she's like, is another form of rehearsing replacements.

One past trainee shared that after they've made an assumption about someone, they use this tool to imagine a short story that contradicts the assumption. For instance, one time they saw a Black man wearing dirty sweats and sitting on the sidewalk; their immediate assumption was that he lived on the street. To contradict that idea, they imagined that he was actually a well-respected professor who went out for a run in his workout sweats, but he injured his leg while running and was sitting on the sidewalk waiting for his spouse to come pick him up. The issue here isn't about which inference is "true"—whether the man was actually unhoused or actually a professor—the issue of interest is **what ideas you want reinforced in your mind** and what mental habits you want to persist.

Again, the goal with this tool is to be intentional about what is rehearsed in your mind. I'm not recommending it as a way to, for instance, encourage someone to ignore the housing problem in their city by imagining every unhoused person they see is actually a professor. Nor should we overapply the rhetoric that "girls are tough" so that we can strip away sensible safety protections for girls or women who might be vulnerable. If you've gotten this far in the book, I doubt you'll be likely to willfully misunderstand things in these ways, but I add this caveat to cover all our bases.

Rehearsing Out Loud

Doing this practice as a mental rehearsal is a good start, and it can be even better if you can do these rehearsals out loud. **We learn even better when we say things out loud.** When you say something out loud, you get the additive effects of first thinking of it, then saying it, and also hearing yourself say it, reinforcing the statement or idea across three modalities. Have a chat with your dog, cat, or houseplant to practice aloud what you want to change!

Rehearsing Behavior

I've mostly talked about rehearsing replacements that are words or phrases, but it can also involve behaviors. For example, a past attendee once shared that she had learned about how people often have an automatic fear or aversion reaction to seeing Black boys and men on the street, as we discussed in chapter 1. Being a member of the Black community herself, she started observing how even members of her own community reacted negatively toward Black boys and men. She even caught herself instinctively frowning when she passed Black boys on the sidewalk. She thought about the consequences of that societal pattern: that Black boys and men go through the world seeing less friendliness. She wanted to work against that, both for her own sons and for the good of her community at large. She made up her mind to smile at every Black boy or man she passes on the street. This story illustrates a different kind of replacement to rehearse. Rather than a thought, it's a behavior—replacing frowns with smiles.

I really liked her solution to this issue, so I adopted it myself. It started off with me putting in the effort to smile at Black boys and men I encountered, and over time it became more and more automatic. Through rehearsal, it became my new default behavior. If I see a Black man or boy, I now smile instinctually. The behavior also started generalizing. At one point, I remember thinking, "So, am I *only* smiling at Black men and boys? What about Black women? Latin folks?" and so on. I started smiling at everyone. I put in the effort to rehearse this behavior because I wanted to work against the specific trend that Black boys likely see fewer friendly faces in the world, but my efforts also generalized to me putting more friendliness out in general. You know what else happens? When you smile at someone, they usually smile back. Their smiles are also new, positive input back into your mind, which can also start working against your biases (as we'll learn more about in chapter 7). If your bias habit is such that you tend to expect less friendliness from Black men you see, the practice of smiling, and its frequent consequence of Black men smiling back at you, works against that habit. You start having more evidence of friendly Black men to work against bias habits that pull you toward thinking of Black men as unfriendly or dangerous.

One time when I shared this "smiling" example in a workplace training session, someone in the audience made a joke about his colleague, saying, "See that, Bob? You'll have to start smiling now!" Everyone in the room chuckled, and I inferred that Bob is probably known for not smiling. I used this joke to reassert my point about how these skills are meant to be customized for each individual person. The smiling worked for *me*. I tend to be an expressive, friendly, smiley person, so putting this replacement into practice fits my personality. I engage in smiling intentionally, but it is still genuine to who I am as a person. If Bob doesn't smile much, then maybe this replacement isn't right for him. We're not talking about putting into practice things that are *completely* out of character or anathema to your personality. Given that Bob's coworkers felt comfortable laughing with him about this idea, I said that I guessed his lack of smiles did not imply that he was unfriendly. So I asked what would be more in line with his personality and suggested that perhaps saying "good morning" was a more genuine expression of friendliness for Bob. And he agreed! Although his demeanor in general was more formal and less "smiley," he typically says "good morning" or "good afternoon" as a polite but friendly way to greet people. For him, therefore, the genuine equivalent of smiling would be to say "good morning" to people he passes on the street. And that's fine! Wonderful, even! We're not trying to change your whole personality here. **We're putting tools in your toolbox, and it's for you to decide what to build with them in your life.**

All these methods of rehearsing replacements (statements, stories, behaviors) can be valid, and what they all have in common is that **they require being intentional in what we want to be practiced and rehearsed in our minds**. Rather than leaving our minds to passively rehearse and repeat what they are fed by our social environments, **we're taking an active role in rehearsing the ideas and images we *want* to have in our minds**.

Rehearsing Replacement Pronouns

One particular area of utility for this tool is when someone comes out as transgender. I have a dear friend, Sandy Eichel, who is gender nonbinary and uses they/them pronouns. I had known Sandy for around ten years before they came out as nonbinary. I therefore had ten years of practice and rehearsal using one set of pronouns to refer to Sandy, and once they came out, I needed to change that verbal and mental habit to use their new pronouns. We all have a lifetime of rehearsal using gender pronouns connected to cues related to biological sex, like body shape or voice, or to stereotypical gender expression, such as clothing and hairstyle. Adjusting your thinking and speech to someone's new pronouns is inherently a process of rehearsing replacements.

After Sandy came out as nonbinary, I started trying to use their correct pronouns, but I would frequently slip up and use their previous pronouns. (Using the incorrect pronouns for someone is often called *misgendering.*) Mistakes and slipups are a normal part of changing a habit; there's no easy "switch" you can flip in your brain to immediately change those well-worn mental associations. We need the practice! But I put in the effort and practice, and over time, as I rehearsed using "they/them" pronouns for Sandy, it got easier and more natural, and I misgendered Sandy less and less often. Practice makes progress!

Correcting Inevitable Mistakes

The topic of misgendering gives us a good place to talk about how to address a mistake you've made. In cases where your mistake involves someone else, don't compound the mistake by making it more of a problem for them. **Just correct yourself, briefly, and move on.** If I ever use the wrong pronouns when in a conversation about Sandy, I briefly apologize, then restate my sentence with the correct pronoun, and move on.

An example of how to do this horribly wrong comes from Sandy's mother's funeral. Before the funeral, Sandy met with the pastor who was giving the eulogy and told the pastor that Sandy uses they/them pronouns. They simply asked the pastor to say that their mother was survived by two "children," rather than saying she was survived by two "daughters." The pastor happily agreed, but when she gave the eulogy, she forgot and referred to Sandy and their sister as "daughters" several times before realizing her mistake. Sandy didn't make a big deal about the mistake, or even say anything about it. But halfway through the eulogy, the pastor realized her mistake, and then went way over the top, making a big scene and saying she was *so* sorry and it's not what she meant and she wants to respect Sandy being nonbinary and on and on—turning Sandy's gender identity into a big kerfuffle mid-eulogy. No one wanted that, least of all Sandy! Because the pastor turned her mistake into a spectacle, much of the rest of people's attention at the funeral became about Sandy and their identity, rather than staying focused on remembering the departed. Sandy wished they'd never even said anything to the pastor. Don't be like that pastor! If you feel the need to have a big emotional reaction to your mistake, wait and do it on your own time! While you're with the person who is the target of your mistake, briefly apologize, correct yourself, and move forward.

A good rule of thumb both for adjusting to new pronouns specifically and for rehearsing replacements in general is this: **When you mess up, take that mistake as a cue that you need to put in some practice.** Generate five or six statements about the person that uses the correct pronouns, or whatever the replacement is that you're rehearsing. Let's say I'm doing this for Sandy. I might practice aloud, talking to my dog or a houseplant, and say something like, "*They* have fun and wacky hairstyles. They are an amazing singer. They have an infectious laugh. I had fun hanging out with them last weekend. I should send a text to their wife and see when we can all get together again." Practice makes progress!

CONSIDER SITUATIONAL EXPLANATIONS FOR BEHAVIOR

Often without even thinking about it, we jump to conclusions about the reasons for other people's behavior. If we see a man trip, we conclude that he's clumsy. If a student fails a test, we tend to call them unintelligent or lazy. If we see a woman crying in public, we label her "overly emotional." When we assume someone's behavior happened because they're clumsy, lazy, or emotional, we tend to see these traits as internal to the person, likely stable over time, and global, affecting many areas of their life. These are what we call *personal* explanations for behavior.

If we put in a little more effort, however, we can think of *situational* explanations for their behavior—things external to the person, malleable over time, or specific to one area of their life. Maybe the man tripped because the floor was uneven, perhaps the student failed the test because they were sick, maybe the woman was crying because someone close to her just passed away. Although we're able to think of situational explanations like these, we tend not to. More often, we quickly jump to personal explanations for others' behavior. To fight back against that tendency, our next tool for retraining your mental processes is to actively **Consider Situational Explanations for Behavior**.

The tendency to jump to personal conclusions is exacerbated when we add bias to the mix. If the student who failed the test was a White man, we will have the general tendency to come to personal explanations as described above, concluding that he didn't study because he's lazy or he's just unintelligent. If the student is Black, however, we still have that general tendency, but then the stereotypes about Black people, which *also* say that Black folks are lazy and unintelligent, reinforce that tendency and pull us even more strongly toward those conclusions about this student.

One of my colleagues saw this particular example play out very dramatically during a training with a group of educators. One of the

CONNECTIONS: PERSONAL EXPLANATIONS AND BIAS HABITS

I also want you to think about how personal explanations for behavior can be exacerbated by the bias habits we learned about earlier. Once you've come to the conclusion that someone is lazy, attentional spotlights will draw your attention more to instances of them being lazy, and confirmation bias will lead you to remember these examples and give them more weight. If the person is part of a group project, maybe you'll assume they were lazy and their group mates did all the work, which becomes an untested assumption. When you give this person a task, your expectation that they won't work hard might lead you to explain the task to them in a more dismissive way, so they don't take it as seriously, becoming a self-fulfilling prophecy. Once we come to a personal conclusion about someone, it can become recursive and strengthen itself over time, making it more and more likely we come to the same conclusion in the future.

teachers, upon hearing this example, disrupted the training session with a loud expletive. She had realized, in the moment, the way bias had played out for her with her students. Whenever she had a White student who wasn't performing well, she'd go and have a talk with them about why—essentially, giving them an opportunity to explain their situation. For her Black and Latin students, however, she would jump to personal conclusions, assuming they were lazy and didn't care about their education. For her, it was a very stark and jarring sudden realization of how bias had unintentionally seeped into her classroom and how she had been systematically disadvantaging many of her students over the years.

Jumping to personal explanations is our brain's quick, default way of coming to an easy conclusion about someone, rather than having to exert energy thinking about them in a more complex, nuanced way. We think we have them figured out, so we don't have to give them any more thought. If, instead, we practice thinking of situational explanations for someone's behavior, it will push back against that easy tendency, which is especially likely to disadvantage members of stereotyped groups. Taking the time to think of situational explanations does not mean we're making excuses for the person's behavior (e.g., "Well, maybe he was sick during the test, so we just won't count it!"). We're not "letting people off the hook" for the behavior, if it is something they need to be accountable for. Instead, the mental exercise of considering situational explanations for behavior helps us slow down and be less likely to jump to conclusions, helping us recognize that we don't really know everything and that there could be so much going on that we aren't aware of.

An attendee of my training program shared a way she used this tool. One day, she received a text message from an employee saying he was going to be late for work that day. She started jumping to personal conclusions about this employee, thinking things like, "He doesn't respect me" and "He doesn't take this job seriously." These thoughts started making her angry, and she was getting "hot under the collar" thinking about him. But then she deployed this tool and started considering situational explanations. Why might he be late for work today? Well, maybe his childcare fell through, so he had to find someone who could watch his kids. Maybe his car broke down, and he had to get it towed, then catch a bus to work. As she generated these situational explanations, she felt herself physically cooling off, having more compassion and being less angry at the employee.

She emailed me to share this story several weeks later, and by that time, she said her entire relationship with that employee had changed and become much more positive. In her email, she never told me what *actually* made that employee late to work that day—and I think that's important to point out. She understood that this tool involves **the mental exercise of generating explanations, not necessarily figuring out what really happened**. In many cases, it doesn't actually matter what the real reason is. Pretty much all our behavior arises from a combination of personal and situational factors. But the *exercise* of generating situational explanations helps slow us down and prevent snap judgments. Another training attendee came up with the mnemonic device **"Get curious, not furious!"** as a way to characterize what happened in this story. When we jump to a personal explanation, we're fooling ourselves into feeling like we have things figured out. Considering situational explanations for behavior helps disrupt this process, maintains our curiosity, and helps us recognize how little we actually know about the reasons behind others' behavior.

DO PERSPECTIVE-TAKING

One of the most powerful ways to help us create meaningful change in our minds and behavior is to **Do Perspective-Taking**. This tool, which you may have heard about before, involves taking a moment to imagine what it would feel like to be in another person's situation. You might imagine what it's like for women doctors to have people constantly question their qualifications because of their gender, what it's like for older people to have others always assume they're feeble, what it's like for a fat person to constantly have people assume they're unhealthy, or what it's like for a bisexual woman to constantly be asked about her husband, when she actually has a wife. **Making the effort to try to understand another person's point of view builds a mental and emotional connection that helps you overcome stereotypes and bias.**

I want to emphasize a key word here: *imagine*. None of us can ever fully know all the nuance and details of being someone of a different race, sexual orientation, or other social group. For instance, as you can see in my author's picture on this book, I unambiguously appear White. I can never truly *know* what it feels like to be a Black man walking down the street and to have people act like they're scared I'm going to rob them. Also, not every Black man will feel the same way about that experience. I can, however, *imagine* what those experiences might feel like. If I walked down the street and people acted scared of me because of my race, I imagine I would feel hurt, frustrated, and angry that I lived in a world where people had those kinds of reactions. When people do this kind of perspective-taking, they tend to think things like, "That's not fair" or "I wouldn't want that to happen to me," which leads them to develop greater empathy and put more effort into working to reduce bias to make sure that they don't contribute to these kinds of experiences.

The Power of Empathy

Doing perspective-taking helps us build a more emotional, empathetic connection to others' experiences. That empathy is important for multiple reasons. By default, we tend to have less empathy for people who are different from us, and we empathize more easily with people more similar to us. Doing perspective-taking bolsters our empathy for those different from us to help contradict this tendency.

Empathy is a powerful influence on our behavior, and reduced empathy can be highly consequential in many domains. If medical professionals have less empathy for a patient, they may give worse care to that patient. If jurors have less empathy for a defendant, they may be more likely to render a guilty verdict. Empathy is a central part of whom we help and support, across many domains and behaviors.

Further, the empathy that we practice when we do perspective-taking builds emotional connections that can have sustained, powerful influences on our behavior. Several years ago, I delivered the bias habit-breaking training as a live, in-person workshop for faculty at a large university. At the end of the workshop, attendees stayed to have additional discussion. One woman in the back of the room raised her hand. She was in a wheelchair, sitting behind the last row of chairs that had been set out for the workshop. She asked, "Did anyone else notice that the way this room was set up doesn't meet accessibility requirements?" Honestly, I had not noticed that. When she pointed it out, I realized that whoever set up the rows of chairs didn't leave enough space for a wheelchair to get down the aisle and didn't leave any areas where someone in a wheelchair could roll up and be included as part of the group sitting in chairs. As we discussed this issue, I learned that this problem was pervasive at her university; she encountered these obstacles daily.

Her experience afforded me an opportunity to do perspective-taking. When I went back to my hotel that evening, I thought back on her experience. What was it like for her to be physically excluded from full participation in the workshop? What did it feel like to have to sit by herself, behind everyone? Anyone wanting to include her in discussion would have to awkwardly bend around the back of their chairs. The situation was made even more palpable by the fact that this was a workshop *about* diversity and inclusion, but she was physically excluded from full participation. That kind of exclusion hurts! The fact that this circumstance *kept* happening to her brought to mind what we learned in chapter 1 about how bias incidents have *cumulative* impacts. I thought about how exhausting that must be for her to face day in and day out. Imagining her experience gave me a real, powerful emotional reaction.

This experience left a mark on me and changed my behavior permanently. I am now unable to walk into a room without looking around to see whether things will be fully accessible to someone in a wheelchair. If chairs or tables are set up in a way that would block someone in a wheelchair, I go and move them to make enough space for people who use wheelchairs or walking assistance devices or who have other accommodation needs. I do this when I am the speaker for an event and also when I am just an attendee. When I first started doing these corrections, it shocked me how often rooms are not set up for people with disabilities, despite federal law in the US—the Americans with Disabilities Act (ADA)—dictating that rooms must be set up to enable people in wheelchairs to fully participate. Very often, rooms get set up by whatever person is willing to do the physical labor of unstacking and arranging chairs—and they don't go through any training or certification to do that work. In my case, however, doing perspective-taking forged an emotional connection in my mind, which then changed my actions going forward, motivating me to put in the effort to make sure rooms I'm in are more accessible.

One additional detail I want to emphasize with this example is that, before that woman spoke up, if you had asked me what the legal requirements were for how a room was supposed to be set up, I could have told you. I had the *knowledge* about how aisles are supposed to have a certain width to accommodate wheelchairs, and so on. The knowledge was in my head, but it was only after doing perspective-taking and connecting that knowledge to a meaningful emotional experience that the knowledge started reliably informing my behavior. It's a common misconception that knowledge in and of itself is sufficient to change behavior. **Emotional connections to knowledge are a powerful way to initiate and maintain meaningful, lasting changes in our behavior.** And that is why doing perspective-taking is such a powerful tool.

Retroactive and Proactive Perspective-Taking

I want you to think of perspective-taking as a tool you can employ both retroactively and proactively. My accessibility example above involved me deploying it retroactively; after I learned how someone was excluded, I thought back on it. I encourage you to do the same, whenever you notice or hear about instances of bias or exclusion.

But you can also employ perspective-taking as a proactive exercise. Work it into your daily routines; for instance, many recipients of this training report that they set aside some mental time to do perspective-taking on their commute to work each morning. Find your way of incorporating perspective-taking into your life.

One way that I do this exercise proactively occurs whenever I fly somewhere. Many years ago, I had a Muslim student who shared that whenever he flies, he always gets to the airport three hours early because of his frequent experience getting pulled aside for additional screening at airport security. I travel all over the US and the world for my work delivering this training program, but I have never once been pulled aside for extra screening. Any time I go through security and it's easy and quick for me, I use that as an external cue to do perspective-taking. I think about what it must be like to get targeted for extra screening so often. How does it feel? How would it change my day and well-being if I had to get to the airport three hours early? How annoying it is for him that he always has to arrive three hours before his flight, just in case he's pulled aside again? Airport security is a context where I know other people face obstacles that I don't have to face, so I practice doing perspective-taking when I am in that context.

CHAPTER 6 SKILLS SUMMARY

In this chapter, we learned about four tools to help retrain your mind!

- The first tool builds on the last unit's awareness skills, as we **Detect, Reflect, and Reject Bias**. This tool involves us detecting and labeling bias in speech, actions, media portrayals, or our thoughts, then reflecting on where the bias comes from, what its consequences are for our own minds or behavior, and what its consequences are for others. Lastly, we want to unemotionally reject the bias as something we don't want in our minds or in the world.
- We also learned to **Rehearse Replacements**, in which we practice thinking, saying, or doing a replacement to our bias habits. We use repetition and rehearsal to bring our habits more in line with our intentions. Practice makes progress!
- We can also **Consider Situational Explanations for Behavior** to slow down our snap judgments and help us recognize how little we actually know about the reasons for others' behavior. Get curious, not furious!
- To help increase our empathy, we can **Do Perspective-Taking** to imagine what others' experiences feel like, forging emotional bonds that help drive our future behavior.

REFLECTION ITEMS

- Recall some of the examples of bias you generated in previous chapters. How might some of this chapter's tools help you retrain your mind to disrupt those biases?
- Identify some areas in your life where you want to start detecting, reflecting, and rejecting bias.
- What are some examples or ideas of replacements you might want to start rehearsing?

- Has anyone ever jumped to a personal explanation for your behavior and overlooked situational explanations? Can you think of times you may have done that to someone else? How could they, or you, use the Consider Situational Explanations tool to stop these assumptions?
- Recognize the difference between *knowing* what it's like to be someone of a different social group versus trying to *imagine* what their experiences are like. Think of your own social identities. What are some aspects of your experiences that might be unique and difficult for members of other groups to truly understand?
- Think of a few scenarios when you might use each of the four tools explained in this chapter to retrain your mind.
- See if you can generate examples that illustrate how this chapter's content is
    - *actionable* (What are you *doing*?)
    - *self-sustainable* (How will you maintain it over time?)
    - *generalizable* (Consider examples related to race, gender, sexual orientation, age, disability, politics, religion, body size, or other groups.)
    - *customizable* (How might you use these skills in your various work, life, and play contexts?)

SHARE YOUR REFLECTIONS!

Your examples and experiences can contribute to our research, help us improve this training program, and provide guidance to others who complete this training in the future! Consider sharing your examples, insights, and ponderings with us at BiasHabit.com/share.

7

TOOLS TO BROADEN YOUR INPUT

A CORE MESSAGE throughout this book is that our thoughts and behaviors are products of our input. Changing our inputs, therefore, can help us change our thoughts and behaviors. This chapter explains several ways to **Broaden Your Input** in service of your efforts to make meaningful change. Expanding the type of information and experiences you have related to stereotyped groups will, over time, make stereotypes less prominent in your mind.

Giving your brain broader, more, and better input related to diverse groups will **push back against preformed biases and stereotypes**. Rather than thinking of a group as just one thing, or one of a few things, you're loading up many different ideas and associations, so **the narrow stereotypes and biases are less dominant in your mind**. You're giving your mind other ideas, emotional connections, and details with which to work.

And, in keeping with our goal of cultivating diverse joy, we hope much of this input will be joyful, entertaining, and fun! We're going to discuss a few major ways you can broaden your input to push back against biases and cultivate diverse joy in your mind.

BROADEN YOUR INPUT VIA GENUINE CONNECTIONS

The most direct way to broaden your input is to **make genuine connections with people from different groups**. When you think about your friends group, the people with whom you interact the most, how many of them are highly similar to you, in terms of race, sexual orientation, and other group statuses? We tend to gravitate more easily toward people who are similar to us on superficial levels. Try to broaden your input by seeking out interactions, friendships, acquaintanceships, and other genuine connections with people who are different from you.

A caveat to mention right off the bat is that we're talking about *genuine* interactions. I'm not saying, for instance, that you should go up to the one Asian person in your work group and ask them, "Hey, can you be my new Asian best friend?" I'm sure you can understand how that's not a nice way to interact with someone. We're not talking about *using* someone else as a tool to work on your biases. What we *are* talking about is **opening up and breaking down barriers that might preclude us from making meaningful connections with people who are different from us**.

Humans like making connections with one another: The key is to overcome the obstacles we ourselves put up that prevent those connections. People often think that they won't have much in common with members of another race, religion, or culture, so they don't even try to connect, which results in us mostly being friends with people similar to us. **Don't let someone's group membership be something that stops you from having meaningful connections with them.**

Just about anyone you meet will have *something* in common with you, regardless of whether they have different identities from you. This maxim is *especially* true when it comes to the things we do for fun. Do you enjoy hiking? Watching sports? Knitting? Video games? Cooking? Horror movies? **For just about *every* person you meet, you can find at least one common interest if you look for it.** I'm sharing this notion as something you can recall later, when you're in a position to forge a new friendship with someone who's different from you. **Approach new people with a positive expectation** that you will likely be able to find something in common. Finding just one small thing in common is the easiest, most straightforward way to start a new friendship with someone.

Let me make this idea more concrete with a story from my life. If you've made use of any of this book's companion videos or podcast, you're already familiar with my dear friend, colleague, and podcast cohost, Dr. Amber Nelson. If we focus on group memberships, she and I couldn't be more different. She's a straight, Christian, Black-appearing woman, whereas I'm a gay, atheist, White-appearing man. She and I met at a conference many years ago. I was invited to this conference as a speaker, and the conference organizers held a social hour for all the speakers and organizers to get to know each other. I was a little anxious coming to this social hour because I didn't know a single person there. But I saw that one of the attendees, Amber, had a Wonder Woman decal attached to her ID lanyard. That decal gave me one easy connection I could make with this stranger. I walked up to her and pointed out the decal and said something along the lines of "I like Wonder Woman too!" That one simple connection was the spark that led to our long-lasting friendship. We started talking about Wonder Woman, then superheroes in general, and the more we talked, the more we found additional connections and shared interests. Now, years later, she's one of my very best friends and one of my closest colleagues.

I've had colleagues tease me about that story and this section of the training, saying that essentially what I'm doing is "teaching people how to make friends." And, yes, I am! It's good to have a reminder sometimes that making a new friend *can* be as simple as finding one small thing in common and chatting with them about it. I think for many adults, we have our established friends circle, and the idea of starting to make new friends can seem daunting. It may feel even more daunting if the potential new friends are of a different social group. But if you remember that all it really takes is finding one small connection—and you believe that you can likely find a connection with almost anyone you encounter—I hope it will seem less daunting.

What does broadening your input by making genuine connections do for you? Well, the entirety of a friendship is input that continually builds new mental and emotional connections in our brains. In the case of my friendship with Amber, our time spent chatting, our inside jokes, and the vacations we've taken together all give me new input related to Amber's social categories, and it does the same for Amber related to my social categories. As one of the important Black women in my life, Amber gives my mind and heart many new connections related to Black women. If I had no such connections with real Black women, all my mind would have is the narrow input provided by media and culture, which more often leans toward biases and stereotypes. My real, meaningful relationships with Amber and other Black women in my life push back against the stereotypes and biases. When I meet a new Black woman, biases will still come to mind to potentially influence me, because we can't completely erase those biases from our minds, but my real relationships and joyful connections with Black women from my life will *also* come to mind, meaning the biases have less primacy and influence.

In my particular life and mind, I think Amber has a bigger influence against my biases related to Christian people rather than Black people. Like many LGBTQ+ people, I've had a lot of exposure to prejudice and hate from Christian-identifying people, both through firsthand experience and through the media, political speech, and protestors.

Whenever I hear that someone is Christian, my "threat antennae" go up. That's a type of bias that I'm keenly aware of in myself. On the one hand, it serves a protective purpose for me; the heightened anxiety cues me in to the potential for someone to express prejudice toward me. On the other hand, I don't *believe* most Christians are hateful, and I certainly don't want to behave in a biased way toward someone just because of their religion.

Anti-Christian bias, or anti-religion bias, is a bias habit I recognize and try to work on within myself. My relationship with Amber is one of the best tools in those efforts. Maybe I catch myself in some impression justification—I hear that Jaden is Christian and start having a bad gut feeling about him because of how my past experiences have built unfavorable associations with Christian folks in my mind. Left unchecked, that bad gut feeling might start coloring whatever else I learn about Jaden. Instead, I recognize that gut feeling for what it is—a bias habit—and I think about Amber. My positive feelings toward my Christian friend come up and start counteracting the negative gut feeling. I can think to myself that maybe Jaden is just like Amber—a Christian who is an enthusiastic ally to the LGBTQ+ community. Bringing up those positive feelings and thoughts doesn't mean those ideas are *true* about Jaden, but it is a way I can use my meaningful friendship with Amber to work against my potential to express bias toward him.

General and Specific Input

I think of the benefits of broadening your input in terms of both general and specific input. Thus far I've emphasized the general input—the increased volume of input, especially positive, friendly input, related to a real person helps push back against biases related to their social groups. All the joy, affection, shared laughter, and time you spend with someone help work against any biases that might relate to their social groups. In my example, these *general* inputs are a happy side effect of my friendship with Amber. I also get more specific input that teaches me new things related to Amber's identities. Maybe she shares a story about a time she experienced discrimination or other ways someone has treated her differently because of her race. These bits of input also serve to help me learn *specific* lessons about how race and bias might impact her life or the lives of other Black folks.

For instance, you may have noticed when I first mentioned Amber, I described her as "Black-appearing." More specifically, Amber is half Black and half Italian. In the mainland US context where she and I live, nearly everyone who sees Amber perceives her unambiguously as a Black woman. A few years back, Amber visited Italy. While there, Amber was surprised that Italian people saw her and recognized her as Italian. In Italy, it was her Italian identity that people primarily perceived when looking at her, which was a palpably different experience for Amber. Amber also loves to vacation in Hawai'i; while there, many people misperceive her as Samoan. When she travels in Mexico or finds herself in other Latin American contexts, people often think she's Puerto Rican or Dominican. As a scientist-practitioner with many years of experience discussing and studying race, I already understood the fundamental idea that race is a social construct: Our notions of race tend to be defined by the culture we are in, rather than major biological reality, and even ostensibly "objective" physical features like skin tone or facial structure are perceived differently based on cultural context. But through this specific input—hearing about my dear friend's real, firsthand experiences with how her race is perceived so very differently in different contexts—I have a much deeper, more human understanding of that idea.

In chapter 3, I mentioned another friend of mine, Nicki Vander Meulen, who is a lawyer specializing in disability activism and a person with disabilities herself. Nicki and I are friends for all the "normal" reasons one is friends with someone else. We like similar movies, we make each other laugh, we have similar values—that kind of stuff. Also, through our friendship I sometimes learn specific content related to disability issues. Sometimes Nicki shares stories of obstacles she had to overcome growing up as an Autistic child with cerebral palsy. Learning about those stories is *both* a part of getting to know my friend's life on a deeper level *and* a tangible education for me about some of the real-world difficulties people with disabilities can face.

Sometimes I learn about the legal cases Nicki is working on (without violating attorney-client privilege, of course!). You probably know that schools are supposed to provide accommodations for kids with disabilities. Very often, however, schools fall short of doing what they are legally required to do, and parents of kids with disabilities have to get a lawyer to force the school to provide the legally mandated accommodations for their kid. Nicki is that lawyer! Learning about these experiences is additional specific input that instills in me more compassion and empathy for those experiences. In fact, maybe I do perspective-taking to think about how it would feel to be in those situations! Learning about these disability issues is a natural result of my friendship, and it helps me understand that the frequency of those experiences is higher than I see in my own life as an able-bodied person. I use all these inputs to motivate myself to keep working against my own biases and in favor of more inclusion for people with disabilities.

Friends as a Resource

Now we've learned about how broadening our input through genuine connections helps us on our journey to build better habits, both through general input that pushes back against biases and through specific input that helps us learn. *After* the genuine friendship is established, your friends can *also* be a resource when you have questions about something relevant to their perspectives or group memberships. My friendships with Amber and Nicki both happened naturally, and those friendships *also* broadened my input to produce beneficial effects. But I didn't set out "trying to find a Black friend" (or Christian friend, or friend with disabilities). I want to be clear that we're not trying to use other people's diversity as a tool to fix ourselves. I've never heard of a genuine friendship built primarily on picking someone just based on a demographic.

The Paradox of Seeking Dissimilar Friends

In some ways, this tool seems somewhat paradoxical—I'm saying both "diversify your friend group" and "don't use someone's diversity to fix yourself." How do we resolve this paradox? Think about genuine friendliness, openness, and being welcoming. Rather than approaching this tool through a solely self-motivated lens ("I'm doing this to work on my bias"), approach it as taking on the goal to **put more friendliness out in the world**, especially toward people who might be less represented within your immediate social circles. Take stock of your friends group, your work besties, and the people you interact with most often. How many of those people are of a different race, nationality, gender, sexual orientation, or other social group status than your own? Now think about the people in your social periphery. Many people find that there is more diversity among their acquaintances, and they can put in some effort to deepen those connections. Cultivate diverse joy!

The Tool in Action

You might also approach this tool in terms of recognizing that some people might be more likely to be left out by others, so you're putting in an effort to include them. I've been in several organizations where I was the only gay man or LGBTQ+ person and I frequently felt left out—not because I think people were *intentionally* leaving me out, but because the more "natural" or easy mechanisms of inclusion operate more strongly for similarity than difference. *Homophily* is a word for this tendency—we gravitate more easily toward people who are similar to us, and before we get to know someone, the basis by which we perceive alleged similarity depends more often on demographics.

Imagine you're in a workshop where you don't know anyone, sitting next to one person similar to you in demographics and one person different from you. When there is a break, which way do you turn to chat? It probably feels a little easier turning to chat with the person who seems more similar to you (this can also be an instance of attentional spotlight!). It's not that you would *always and only* turn in that direction, but talking to a stranger who appears superficially similar probably feels a bit easier than talking to a dissimilar stranger. These are some of the mechanisms that create bias, because we end up having more experiences and familiarity with people who are similar to us. **Broadening your input via genuine connections corrects that bias by cultivating diverse joy!**

Whom do you invite to things? Don't let group membership be an obstacle to including someone. Let's say Jen and Rob are throwing a dinner party. They have six friends in mind to invite, but they could fit two more at their dining table. As they discuss whether to stick with their current list or expand it to invite two more, Rob notes that everyone they're planning to invite is White, and he asks whether they might have any friends of color to invite. They think of an Asian American couple they met a few weeks ago, but Jen muses, "They probably don't want to hang out with a bunch of White people," so they don't reach out. Jen let her (untested) assumption create a barrier to inclusion. Maybe that couple *wouldn't* want to come to the dinner party, but Jen made the decision for them, precluding the possibility that they would have loved to attend!

If you *ever* notice yourself saying or thinking that someone wouldn't want to connect with you because of a difference in group membership, I strongly encourage you to stop yourself. Choose to try! Let the other person decide if they want to engage or not. Do the friendlier, more welcoming thing.

A past attendee of our training shared how he took this advice to heart and put this tool into practice. Mark and his family are White and live in a predominantly White neighborhood. When a Mexican family moved into a house down the street, Mark considered taking his family to go meet them and caught himself having the thought that "they don't want some *White people* coming to bother them!" He almost let that thought be the end of things, but then in a flash, he remembered that I had *specifically* warned against that kind of assumption. As he thought about it more, he remembered how he'd taken his family to welcome *every* new White family who'd moved into his neighborhood. If he'd had any concerns about "bothering" those White families, they didn't stop him. So he detected/reflected/rejected that thought, then gathered his wife and kids to go welcome the new family. It turned out his family and the new neighbor family had lots in common, and their families became very close friends with one another. Also, Mark's family was the first to welcome the new neighbors, and once others in the neighborhood saw his family go over there, they followed suit. Once he overrode his inclination to *not* be inclusive, it set off a chain reaction of welcoming gestures from others in the neighborhood. Perhaps all his other neighbors were having similar thoughts that the new family wouldn't want to be bothered, but seeing Mark's family make the effort encouraged them to override those thoughts as well. In this case, **the inclusive, welcoming gesture was contagious**.

Recall my earlier example about Jake, who didn't invite his coworker Kevin to his Super Bowl party because he assumed Kevin didn't like football because he was gay. If Jake is working on disrupting bias habits, he would hopefully catch himself making that untested assumption (maybe he'd detect/reflect/reject it!), and then reach out and invite Kevin to the Super Bowl party. It's possible that Kevin actually likes football! Or maybe Kevin doesn't, but his husband does, and so they'd still like to attend as a couple. Or maybe he doesn't like football, but would still like to come and be included to enjoy the food and social time hanging out with his coworkers. Maybe Kevin comes to Jake's party out of politeness, but once there, discovers that he and Jake have similar tastes in video games, and they're able to connect over that. Perhaps Kevin can't or doesn't want to attend the party, but he appreciated being invited, then maybe asks Jake out to coffee the following week, and they start building a friendship that way. ***Not* inviting someone only has one outcome—they're not included.** Inviting someone could lead to many different positive outcomes in which you might connect meaningfully.

Acquaintances Count Too!

Although my earlier personal examples emphasized very close friendships in my life, I'm not saying you have to go out and become best friends with everyone. Not everyone you engage with will end up being a super close friend, and that's okay. Having a pleasant acquaintanceship is also nice. Maybe the "one small thing" you find in common with a coworker is that you both love hiking, so they become primarily your "hiking friend." Sometimes you go hiking together, and other times when you go hiking without them, you come back and mention it to them the next time you see them. And you continue to have short chats about hiking, why you like it, and so on. Maybe the acquaintanceship never gets much deeper than that, but it's still a nice piece of genuine friendliness you're putting out to them, and you've broadened your input and cultivated diverse joy in a way that will do some work against bias habits in your own mind.

Make genuine connections with different people to counteract any tendencies toward homophily within yourself. You can also employ this tool as a way to correct tendencies outside of yourself. Recall my example in chapter 6 about smiling. That involved recognizing a potential pattern of unfriendliness in the world and correcting it by increasing the friendliness you put out in the world. As I noted earlier, that practice *also* leads to broadened input: After you smile at folks who are different from you, when they smile back, it's positive input for your brain to build more joyful associations with that social group.

In addition to your efforts with people in your typical social orbit, you can also think about how to **make genuine connections with people in other cultures through travel**! Mark Twain wrote about how travel can be used to broaden your input, saying, "Travel is fatal to prejudice, bigotry, and narrow-mindedness, and many of our people need it sorely on these accounts. Broad, wholesome, charitable views of [people] and things cannot be acquired by vegetating in one little corner of the earth all one's lifetime." I think travel is an excellent way to broaden input, but you can also find many opportunities for additional connections closer to home. Wherever you live, look out for various community events or organizations that bring together different types of people. Many LGBTQ+ Pride organizations, for example, put on events for the express purpose of promoting goodwill between straight and LGBTQ+ people. There are countless similar opportunities—cultural festivals, speaking engagements, community outreach events, and so on—if you look out for them. They happen all the time.

Be Thoughtful

One concern that sometimes arises is that this tool might lead to "singling people out" and adding a burden on someone who is already disadvantaged by bias. Let's say there's a work group of twenty people, and Amy is the one person who is different from everyone else on obvious demographic features (maybe she's the only person of color, the only woman, or the only LGBTQ+ person). The work group reads this book as a diversity activity. A concern might be that Amy's nineteen coworkers will all try to be more social with her, adding an unfair burden on her time and energy, *on top of* the fact that she's already the most likely person to be disadvantaged by bias in this workplace.

First off, I want to acknowledge that this concern is valid, and it's good for us to think about it. I am a very social person, and even I feel a little stressed out if I imagine being in Amy's shoes and getting nineteen invitations to hang out all at once. But I don't think that possibility is very likely. It implies that *none* of Amy's coworkers has ever been friendly with her before, and they're all *just* reaching out to her because a book told them to. Frankly, if the real example is that extreme—if *no one* has ever gone out to coffee with Amy—I think that drastic level of exclusion would the bigger problem, not the efforts at inclusion that people start making.

What is *more* likely and realistic? Everyone will have different levels of responsiveness to the tools in this book. There might be *one* of Amy's coworkers who realizes they've never had coffee with her and texts her as soon as they read this chapter. Three or four people might think about Amy, then realize they already know her pretty well, and so they use this tool to focus on cultivating genuine connections with someone else. One person thinks of Amy a month from now and invites her to their party. And so on. Just **make sure that, however you reach out, you're doing it out of genuine friendliness**. And respect the other person's response. Maybe Amy isn't interested in connecting with you. That's her choice! You're making yourself open to building new connections, not forcing them on others. You're offering an invitation, not imposing an obligation.

One supervisor in another workplace employed this tool by resurrecting a tradition his supervisor used that made things fair for all his employees. Rather than singling out any one employee, he started getting coffee one-on-one with a different employee every week. That way, he gradually worked his way through all his employees, and each of them got an equal opportunity to connect with their supervisor. This approach also helped him make sure that he was reaching beyond standard demographics like race or gender; after all, maybe his particular habit is to gravitate toward employees who are more extroverted or who have a similar educational background to him. By employing his version of this tool to connect with everyone equally, he made sure no one was left out.

BROADEN YOUR INPUT VIA MEDIA

As we learned in chapter 1, the media is a major culprit in teaching and reinforcing biases. If, however, we can **be intentional about the media we consume**, it can also be a powerful tool against bias. Broaden your input by seeking out books, movies, television shows, social media accounts, podcasts, blogs, or other media that have been created by people whose backgrounds are different from your own or who tell the stories of people different from you.

When I suggest using the media to broaden your input, very often the first type of media that comes to mind is educational content. People think of historical biopics or documentaries or books written by luminaries and thought leaders who reflect on their people's experiences in modern society. Let me start by saying, yes! Please do engage with those and similar forms of media! I think their value is self-apparent. I'm not going to talk at length about educational content, both because its value is self-apparent and because plenty of other people, including the creators of those media items, already advocate effectively for these stories. For example, one of the trailers for Netflix's 2023 historical drama *Rustin* starts off saying, "You may not know the name Bayard Rustin, but you should!" Like most educational media, this movie quite literally speaks for itself. It tells you there is an important historical figure most people haven't heard of, but the creators think you should learn his story! I believe learning is inherently valuable, and learning about the historical and contemporary lives and experiences of people different from you is one excellent way to broaden your input.

But the key to this chapter's tools is to *broaden* your input. We're going to emphasize engagement with educational content *and* entertaining content! Think also about comedies, musicals, murder mysteries, short stories, poetry, science fiction! Media that brings you joy! If you *only* engage with explicitly educational media, your input remains relatively narrow. In fact, having *too* narrow a focus on educational media can sometimes lead people astray, and before I move to entertainment media, I'd like to caution you to watch out for four pitfalls that can sometimes happen with educational media.

The Pitfalls of Educational Media

First, **don't let your input be *so* uniformly sad that it demotivates you**. I once met a woman who was a part of an "anti-racism book club for White women," and every book they read dealt with historical or contemporary atrocities and difficulties faced by Black people in the United States. After a while, she realized that whenever she thought about Black people, she just felt sad. Women started dropping out of the group—some even saying that it was because it just made them sad all the time. I see variations on this pattern a lot. People focus too narrowly on the atrocities a group has faced, and it leads them to feel sad, perhaps guilty and shameful, and, ultimately, powerless. The issues of racism or other forms of bigotry seemed too huge and pervasive to fix. The reading group members felt powerless and gave up trying to make any meaningful change. For people falling into this pitfall, the net effect is that they become *less* motivated and less engaged with making any sort of positive change.

Second, **don't use others' trauma as catharsis**. I occasionally encounter people who see "feeling bad" as an end in itself. In other words, they've read or watched or experienced something that made them feel really bad, and they see that as a sort of punishment to alleviate their guilt. Spending some time feeling very bad about racism becomes a figurative self-flagellation to pay for the self-assessed "sin" of being White. Don't be a trauma tourist! The most extreme example of this catharsis pitfall I've ever heard of was some White folks who paid to have people of color yell at them and tell them how awful and racist they were. It is unclear to me how such activities help us make meaningful progress on real-world problems. A piece of educational media may have a net effect of making you "feel bad," as you have compassion or empathy for atrocities in history or difficulties in the modern day, and there's nothing wrong with that. Just don't use that as an end in itself. See it as a means to motivate yourself to put more work into making positive personal and social progress. Time spent "feeling bad" doesn't let us off the hook; it doesn't absolve us from having to put in real-world effort.

Third, **don't use contemporary or historical extremes as a basis for positively judging your own behavior**. Sometimes people focus on the horrendous chapters of human history, such as the Nazi Holocaust, slavery, genocides of Indigenous people, or hate crimes, and (consciously or nonconsciously) use them as a barometer to judge their own behavior favorably. For instance, let's say Jaime watches the PBS documentary *Rising Against Asian Hate: One Day in March*, about the rise in hate crimes directed at Asian and Asian American people during the COVID-19 pandemic. He thinks about how racist and bigoted the people committing the hate crimes are and how he would *never* commit such heinous acts. Jaime feels pretty good about himself because he is far less awful than those hate criminals. This extreme downward social comparison leads Jaime to be less motivated to work on himself. This pitfall leads people to become less motivated to make progress because they're comparing themselves to an extreme.

Sometimes we call this phenomenon *anchoring on the past*: When compared to the lives of enslaved Black people before the US Civil War, the lives of Black Americans today seem pretty great! People say or think, "Look how far we've come!" and figuratively pat themselves on the back, overlooking actual problems modern Black folks face. The end result is less motivation to work on things in the present. Let me say again that I *do* strongly encourage you to engage with educational content, even content focused on atrocities! It's good to learn about the injustices of the past and the present, and also, it can be good to recognize what progress has been made. Just make sure the end result is *more* motivation to make progress, rather than *less*.

The fourth and final pitfall I want to discuss involves media that has more of an uplifting emotional impact. Rather than focus on media that revels in tragedy, some people narrowly focus on inspirational figures, with the unintended pitfall of having less concern about bias. For instance, perhaps someone watches a series of lighthearted documentaries about the accomplishments of inspiring Latin Americans, but never engages with any content that discusses the barriers or hardships Latin Americans might face. Or maybe someone making this misstep engages with media about successful Black people, then decides that "racism isn't a pervasive problem" because some Black people have great success.

A common cultural term that relates to this pitfall is ***toxic positivity*—only seeing the good and optimistic messages without also engaging with difficult or negative realities**. Disability advocate Stella Young named a version of this phenomenon "inspiration porn," in which nondisabled people gush over how inspirational it is when someone with a disability accomplishes a relatively mundane task. The most extreme examples of disability inspiration porn have no educational value at all, but other content might be educationally valuable up until someone *uses* it as inspiration porn. As with all the tools, *intentionality* matters. Are you consuming content *just* to make yourself feel better, or are you actually learning something meaningful and broadening your input?

You've probably gathered by now that all four pitfalls involve people ultimately becoming *less* engaged and *less* motivated to work on creating meaningful change or building a better world. Remember our model for change: Motivation is the first necessary condition! We want to maintain and elevate our motivation to make progress, not snuff it out! Especially when we're talking about educational media, I want you to keep in mind that we're using this tool *in service of our goal to be ongoing, effective agents of change.* As long as you remember that, I'm sure you won't fall into any of these pitfalls.

Educational Versus Entertainment Media

Earlier, when I drew a distinction between educational and entertainment media, I did not mean to imply that they are fundamentally distinct and mutually exclusive. I think good educational media is engaging (i.e., entertaining), and quality entertainment is an art form that teaches us important lessons about life. As people more poetic than I have said, "Art educates the soul."

The rhetorical distinction I'm drawing between educational and entertainment media is more connected to how people approach the media in question. When you're about to engage with a new book, movie, or television show, are you approaching it with solemnity and seriousness or with joy and delight? Loosely defined for the present conversation, "educational" content will more often lean into seriousness, and "entertainment" will lean more into joy. Because the various forms of prejudice, bigotry, inequity, and bias are serious issues, most people want to take them seriously. And we should! It does not follow, however, that all our efforts to create meaningful change must be joyless. We need *both* types of content in our media diets! **Serious contemplation *and* joyful romps both help us combat bias and create lasting change.** It's the salty *and* the sweet (and the sour)! With regard to broadening your input via the media as a tool to disrupt bias habits, I think people, on average, rely more heavily, or exclusively, on the educational, serious types of media. Remember that on this journey we also have another, intertwined goal of cultivating diverse joy! Without shooing you away from educational media, I want to emphasize the power of joyful, entertainment media.

Consume Broadly and Joyfully

What sort of media do you enjoy? Murder mystery novels? Horror movies? Buddy comedies? Soap operas? Whatever comes to mind, or whatever you're in the mood for, start there! Pick something from a category you intrinsically enjoy but that is told from a different perspective or by a different creator than the media you usually consume. A simple internet search for "[genre] [media type] by [demographic]" will come up with many options! For example, if you love horror movies but notice that most of what you watch happens to be from White creators, a search for "horror movies by Black creators" will come up with various films, including many of Jordan Peele's excellent horror films! If you love Nicholas Sparks romance novels, try some by Ann Liang.

I don't presume you *only* engage with media created by members of your own identity groups, but by default, the decision to consume a piece of media is often easiest if we readily perceive a similarity between ourselves and the characters. Just as we discussed earlier with friendship, people often initially think they won't easily connect with media from a different perspective, but when they are open to it, they are surprised at how much they find to enjoy.

Just like friends in the real world, the characters in media we consume help us broaden our input. Media that means something to us becomes part of who we are. We form emotional connections with characters, learn to see and feel things from their perspective, and come to understand their likes and dislikes, their strengths and foibles. We find both things we have in common with them and ways we see the world differently. All this input is valuable content to put in our minds and hearts that can push back against stereotypes and biases.

Many media companies and their employees have come to recognize this tool as important, often joining the common rallying cry that "representation matters!" As such, many media companies provide ways to make it easier for us to broaden our input. Many television streaming services now have categories devoted to media starring or created by members of historically underrepresented groups. These categories are often most emphasized seasonally during months that the US traditionally recognizes as heritage, pride, or history months, but they exist year-round. For instance, in the US, May is both Military Appreciation Month and Asian American and Pacific Islander (AAPI) Heritage Month. When I open streaming service applications in May, they emphasize and advertise their collections of "Military Stories" and "AAPI Stories." Make use of these opportunities! One of the reasons heritage months exist is to draw more attention to stories from those communities. I want you to apply these tools year-round, but you can *also* take note of the various heritage months and use those as cues to encourage yourself to broaden your input related to a particular group.

Just as we discussed with friendship, broadening your input via media creates benefits both in terms of general associations and specific things you learn. In other words, merely getting *more* input helps push back against the narrow stereotypes in your mind in general. Also, you will sometimes learn specific lessons; maybe a Jewish character on a television show faces hostility at work when they need time off for a religious holiday—if you've never faced that sort of problem yourself, this show helps you start to understand that experience. Or perhaps two Indian American characters each discuss their different feelings related to the Hindu holiday Diwali, so you learn something about that particular religious and cultural holiday and about at least two of the varied perspectives different Indian Americans might have on that part of their culture. And so on.

Although I'm encouraging you to diversify your media diet, that doesn't mean that each piece of media is "diverse" itself. Let me tell you about a sitcom called *Grand Crew,* which was created by Phil Augusta Jackson and aired on NBC for two seasons. It's about a Black group of friends, their lives, and the often very silly shenanigans they get up to. I think it's a great sitcom, for all the typical reasons one might enjoy a sitcom; the actors are all excellent, the jokes are funny, the characters are relatable, and the storytelling is crisp and easy to follow. If we assess this show in terms of a checklist definition of "diversity," it's not very diverse: All the main characters are Black. Within that identity group, however, the characters are all well written, which of course means they are considerably diverse in terms of their personalities, professions, interests, and so on. As you get to know each unique character living their life on the show, all that exposure is input to help push back against stereotypes or other monolithic ideas about Black Americans. It gives us exposure and input related to a variety of Black experiences. I think *Grand Crew* is a great piece of media to bring some Black joy into your life, as one part of your efforts to cultivate diverse joy across your media diet.

I think actors can play a special role in this tool. Think about the input your brain receives as you follow a talented actor across many different roles and characters. If you watch Sandra Oh as a competent surgeon in *Grey's Anatomy* and as an irresponsible party girl in the movie *Quiz Lady* and as an obsessive spy in *Killing Eve,* and *many* other characters in other movies and shows, your brain gets accustomed to that one person occupying many different roles. Sandra Oh is just one particular Korean-Canadian-American actor, but through her work, you come to associate many different traits, ideas, and emotions with her. If you can see one Asian person in so many different ways, it's harder to turn around and think other Asian folks all just have one stereotypical set of traits.

In addition to traditional media like movies, books, and television, you can broaden your input with newer forms of media as well. Even social media! How many of the accounts you follow are from people different from you? Follow some social media creators who are different from you but who make content in a domain that you intrinsically enjoy. You can also share their content, or any media you enjoy, with others!

Podcasts are another form of newer media that can be an excellent way to broaden your input. Whenever I mention podcasts now, I worry people will think I'm only bringing them up as a way to encourage you to listen to my nonprofit's podcast, *Diverse Joy*. I do want to encourage you to check it out; it was designed to accompany this training program, and with this tool in mind specifically, to give you both joyful and educational content, and every episode ends with a media recommendation to help you broaden your input! But in fact, I emphasized podcasts as an especially powerful way to broaden your input for many years before we started *Diverse Joy*. One of the nice things about podcasts is that they are relatively inexpensive to make compared to television shows and movies, which means if you look for them, you can find an impressive range of diverse voices in the podcast ecosystem. Look for podcasts that are related to topics you intrinsically enjoy or things related to your hobbies or passions, but from the perspective of someone different from you.

Recall the earlier story about Rebecca, who identified that she disliked women sportscasters because of norm enforcement. One way she could combat this bias habit would be to seek out podcasts where women talk about sports (a quick internet search for "sports podcasts by women" brings up a lot of them). Because of tacit norms that say men (not women) should talk about sports, and Rebecca's greater familiarity with men talking about sports, the women podcasting about sports might annoy Rebecca at first, continuing the biased feeling she had previously recognized. But the podcast gives her more and more exposure to this input, gradually making "women talking about sports" seem more normal and familiar to her, and less annoying. In this way, she will have broadened her input to address a specific issue she had identified in herself.

Similar to Rebecca, you can deploy this tool in service of specific goals or potential problem areas you recognize for yourself. A probation officer who attended one of my trainings told me that he was *especially* worried about biases related to race and crime. By definition, all his cases—the people he worked with as a probation officer—were people who had been convicted of crimes. These folks were of all different races, of course, but he could tell that when his cases happened to be Black folks, they served as pieces of confirmatory evidence related to racial stereotypes. He was especially worried about how this daily input, exacerbated by confirmation bias processes, likely strengthens those stereotypes in his mind more than the average person. Having recognized a specific problem area for what is most reinforced in his mind, he could then cater his media input to work against that problem. He could put in effort to consume media that portrays Black folks in situations unrelated to or even the opposite of crime and violence.

When you find a movie, book, podcast, television show, or other media you like, tell others about it! The hope in this section is that you'll find lots of media you genuinely enjoy from creators of all different backgrounds. Share that diverse joy with others! If you genuinely like a movie, many of your friends and colleagues probably will as well. People often overlook media from people different from themselves, thinking it wasn't made with them in mind. And maybe it wasn't! Even if you weren't the target audience, it still might delight and enlighten you!

BROADEN YOUR INPUT VIA PHYSICAL, DIGITAL, AND SOCIAL REPRESENTATION

Another way to broaden your input is by increasing representation of underrepresented groups in our environments. Whether we notice or not, the world around us is full of messages, or images that relay messages, about who belongs and whose perspectives are important. These messages can come from many sources—pictures on our websites, portraits displayed in important buildings, and even the art we display in our homes. Take stock of images around you and what groups and perspectives are represented, and **see what you can do to increase**

representation of underrepresented groups. We can increase these representations via aspects of our environments that are physical (e.g., art, photos on our walls), digital (e.g., screen savers, pictures on our websites), or social (e.g., speakers we invite to a club, authors on a course syllabus, organizations we partner with).

One of the most iconic examples of this phenomenon is when organizations have portraits of past leaders or retired board members on display in their conference rooms, lobbies, or other prominent locations. Very often, organizations lack diversity in their historical membership or leadership, which results in these displays being populated exclusively or primarily by White men. For the contemporary members of the organization, these displays often make women and people of color feel as though they don't belong. The unspoken, and probably unintended, message conveyed is that "to succeed here, you must look like these people!" It can affect clients or potential clients as well; a hospital I worked with identified that their "wall of past presidents" in their lobby exacerbated problems with Black patients who were already inclined to mistrust the White medical establishment.

I'm not saying organizations should ignore or erase their histories or that you should go around tearing down all the pictures of White men. But when you look around at whose images are given prominence, think about how you can increase representation in a way that represents your values and goals. Maybe that hospital, for example, could display portraits of Black community leaders who have worked with the hospital, or a Filipino doctor who developed a new technique at the hospital, to start increasing representation in their physical environment.

One academic department I worked with addressed this issue by expanding their portrait wall, so that instead of just honoring their retired professors (who were mostly White men), they added portraits of all the current professors as well, which brought in many more women and people of color. A company I worked with decided that their "wall of past CEOs" didn't really match their organizational values, and they replaced it with a display that instead featured the kinds of customers and clients whose lives their work impacts. Another group moved their "history wall" online to a section of their website that rotated through people one at a time to make it less overwhelming than the physical wall of portraits had felt.

Another academic department chose to take down their wall of portraits and replace it with a display that changed every month. Half the display was used to specifically highlight diverse voices in their field, showcasing scientists from historically underrepresented groups and their work. The other half of the display was devoted to highlighting the accomplishments of one of their past professors, showing not only their picture, but also describing their important research accomplishments. They felt that this approach actually gave *more* respect to the past professors, because before, when it was just a wall of portraits, there was no mention of the research the professors did, leaving their lives' work unmentioned. The new display, in contrast, actually honored their work.

A university I worked with installed an expensive mural in a new science institute they built. The mural was designed to showcase many of the important scientists and contributions that had come out of this university. After it was unveiled, however, people pointed out that the mural only showcased scientists who were White men (perhaps as the result of attentional spotlight?). The university decided to remake part of the mural, at a considerable additional expense, to *also* highlight women and people of color who had made important contributions to their areas of science.

Sometimes applying this tool may be a part of something grand like a giant mural, but much more often, it can be something mundane and everyday. Whose pictures are shown on your website? In brochures or flyers you make? On your social media? In one training I gave, a scientist in the audience told me he runs a social media account that puts out facts about his area of science. The profile picture of the account was always a famous scientist from the field, but he realized it had been only White men, so he picked a new profile photo to highlight a prominent woman in his field. **Find *your* ways of increasing representation.**

One of my colleagues, Dr. Mahzarin Banaji, suggests employing this tool using computer screen savers. Let's say we're trying to broaden our input about scientists who are women of color. We could set up a computer screen saver that is a slideshow of women of color scientists, with pictures of them and some facts about their work or their life. Every time the computer goes to sleep, we can then glance over at it and be reminded of those women and the impact they've made. It becomes a frequent dosage of stereotype-disconfirming evidence! **Broadening your input is a direct method to try to disrupt the effects of attentional spotlight and confirmation bias** we learned about in chapter 4. By giving ourselves a higher dosage of stereotype-disconfirming information, we work against the tendencies to give disconfirmatory information less attention and less weight in memory.

Beware Missteps

As with everything in our tool kit, apply this tool thoughtfully and intentionally. **Don't use images to fake diversity you lack.** For instance, I've seen schools or companies that want to appear to have more racial diversity than they actually have, so they use photo editing or stock photos to misrepresent themselves in advertising or recruitment materials. I do not recommend that approach. For starters, it's lying and false advertising. If you are trying to recruit a more diverse workforce or client base or student population, deceiving prospective recruits will just backfire when they see that they were misled and the actual makeup is not what was shown.

I'm not saying you can never use stock photography; just don't use it to lie. Ask yourself whether you're using stock photography to hide something or because it's just what you would normally do. If you are doing a presentation and need a picture of some doctors, a stock photo of a diverse group of doctors is probably fine! If you're trying to imply that the pictured doctors are *your* medical staff, then a stock photo is not the way to go. The same goes with photo editing. If Joe was absent on the day you took your staff picture, but you want him included, editing him into the group picture—with his consent—is probably fine, provided you're doing that for anyone who was absent, not just the people with identities that bring in visual diversity. Be sensible and honest about it! How you represent the diversity of your organization is *not* a place to "fake it until you make it."

Another potential misstep many organizations make in this arena is when they only have one or two "visibly diverse" folks, and in an effort to showcase diversity in the organization, those people get shoehorned into every picture. It's a form of objectification; please don't *use* someone in that way. These issues aren't simple; it makes sense that you might want to showcase the diversity you have. But rather than unilaterally using someone in service of your goals to showcase diversity, talk with them! Discuss your organizational goals related to diversity and representation, and see if they are willing to *partner* with you in service of those goals. **Respect them and their agency, and proceed only with their consent.**

Seek Diverse Partnerships

If your organization is currently lacking in terms of its own diversity, rather than faking it, one way to get started changing that is to seek out new partnerships. Where could you bring in people of different backgrounds? Maybe you can find diverse voices to bring in as motivational speakers, team building trainers, or representatives from other locations of your organization. Maybe you can partner with community organizations to put on events that appeal to a more diverse group of people. Most communities in the US have organizations that work with underrepresented folks, such as chambers of commerce devoted to particular groups (e.g., Black, Hispanic, LGBTQ+ Chambers of

Commerce). Partner with some of those to put on an event that will bring together your organization and members of that community (this is also another way to work on broadening your input via genuine connections). Maybe you'll find new contractors, new customers, or new employees. At the very least, you're increasing representation in your social environment and you're putting in effort to work with those diverse communities in a genuine way. (And, also, maybe you'll get some pictures at the event you can use to showcase *work* you do that serves diverse communities, even if you can't showcase diversity *in* your organization. I say this in a parenthetical because, of course, we don't want to do this sort of outreach *just* for a photo opportunity. Do it because it matches your organizational values and goals, and the photo opportunity is a fringe benefit.)

Think Beyond Photography

Thus far, most of the physical and digital representations I've spoken about are pictures of actual people, but you can also broaden your input via things like art, slogans, values statements, and movements you believe in. You can display buttons or stickers that remind you about diversity-related causes you support.

These representations can be cues that remind you to put in some practice on these tools. Maybe you put an LGBTQ+ affirming sticker on your car, and every time you see it, you do some perspective-taking or practice rehearsing a replacement. You can even use this book: If you store it on a shelf where you see it every once in a while, it can be a cue to remind yourself to work on disrupting bias habits and cultivating diverse joy! By broadening your input using pictures, art, slogans, or other reminders about your commitment to overcoming bias habits, you're giving yourself cues to bring this work back to the forefront of your mind.

These can also be methods for an organization to start working on things when it doesn't yet have the actual representation it wants. Rather than "faking" diversity in images, like we discussed earlier, you can display something that voices your personal or organizational values related to diversity. We don't want these to be empty words, of course—**they should be words that you and your organization back up with actions**. However, the words can be a start! Let's say you and your organization decide that your organizational values include "advocating for social justice" or "serving diverse communities" or "amplifying diverse voices" or whatever would be genuine and actionable for your group. Then you put those values statements up in the front lobby of your building, prominently displayed, where everyone coming to work sees them as they start their day and again when they finish their day. Each time each employee sees and reads those values, it's a bit of input reminding them what your organization stands for. And if or when you're falling short of those values, they provide a basis for someone to speak up and to point out where you're falling short. **Again—we never want these to be *just* words—but words are often the start, and then we hold ourselves accountable to translate those words into actions.**

If you're an educator, look at whose perspectives are represented in your syllabi or course materials. Many educators and schools I've worked with have done surveys of how many readings come from people of color, LGBTQ+ folks, non-Christian folks, and so on, and that's often one of the most direct ways to see where one might increase representation. One professor told me how he made author representation a new assignment for his class. Every semester, he asks his students to seek out research papers that would fit into the topics of his course, but from perspectives that are underrepresented. This activity serves one of his course's learning objectives—teaching students how to do literature searches—and also gives him an opportunity to talk about diversity and representation. Through this assignment, he's been able to increase representation in the readings in his course.

Whenever I discuss this tool, I think people gravitate more readily to pictures of real people and less readily to art. But I always like to emphasize art especially. I have a friend who is a world-traveling art collector. If you ask him about any of the pieces of art in his home, he'll tell you all about the life story and cultural context of the artist. I remember one time I liked one of his new pieces, and he told me all about the artist, who was a young gay man in Mexico who had to overcome many different sorts of hardships on his journey to become a successful artist. The artist's life story was as palpable for my friend as the art itself. The art my friend surrounds himself with, therefore, is one way he broadens his input. When he sees a piece, it brings him joy and also reminds him of the artist. Maybe he uses the art as a cue to do some perspective-taking, to think about how hard the artist worked to overcome hardships. My friend might also remember the trip he took when he bought the art and the things he learned about the culture while he was there. Even if a piece of art doesn't depict *people*, you can use the artwork as a cue to bring to mind the artist's life, their culture, or your own knowledge and experiences about their culture.

Outward Benefits

I've mostly discussed this tool in terms of its work on pushing back against bias in your own mind. That's definitely where I want us to start, but when we're talking about images and slogans that are public facing, this work can also have benefits for others who are seeing members of their group represented. As a gay man, when I see companies use images of same-sex couples in their advertising, it makes me feel good! It makes me think that company cares about the perspectives of people like me, and I feel greater affinity and belonging related to that company.

Of course, I'm not naive—I also know that corporations are, at least in part, showing me what I want to see in order to get my money. So I try to be diligent and look into whether those companies actually have beneficial policies for LGBTQ+ people behind the scenes as well. But, as people say, **representation matters**! This public-facing benefit applies to your individual efforts as well. If, for instance, someone displays an "LGBTQ+ ally" sticker on their office door, LGBTQ+ people seeing that may feel safer and greater belonging. And then, if, perhaps, someone faces a problem or hardship related to their LGBTQ+ identity, maybe they see the person displaying that sticker as someone safe to talk to.

The other ways of broadening your input also have outward-facing benefits. As you broaden your input via genuine connections, you're putting more friendliness, inclusion, and belonging out in the world as you become more open and inviting to more people. As you broaden your input via the media, the money you spend and the downloads, views, and other metrics media companies use will show an increase in people's interest in diverse stories, making it more likely they fund projects like those in the future. Although we're starting our focus on how these tools can disrupt bias and cultivate diverse joy in terms of creating *personal* change within your own mind and life, you can see how they also start contributing to *social* change and building a better world around you.

LISTEN WITH HUMILITY AND CURIOSITY

When you're broadening your input, you'll find both similarities and differences with others' life experiences, perspectives, and ways of seeing the world. **Similarities *and* differences are potential sources of joy and learning.** Especially when you encounter differences, I want to encourage you to **Listen with Humility and Curiosity**. In fact, try to approach all new input with a spirit of humility—part of why you're broadening your input is because you recognize that your perspectives and life experiences have limits—and with curiosity—when you're exposed to something new or different, choose to be interested in it and learn more about it.

Remember from chapter 2 that when we encounter something unfamiliar, especially something that might contradict our typical way of thinking, we often feel frustrated or annoyed, at least initially. Being exposed to new and different perspectives while broadening your input is one area that could bring up these frustrated feelings. When you learn that someone sees the world or a situation in a fundamentally different way from how you see it, it can feel like a challenge—like they are saying your perspective is wrong. If you approach things with humility and curiosity, that can help disarm these reactions, as you seek to learn more about how the other person sees things.

This tool is especially relevant when someone else points out that something (a word, phrase, or action) is offensive or problematic in some way, but you don't understand why. If someone says that something is offensive, whether it's someone in media you're watching or someone you're talking to, be curious about it! If you don't understand why that person found a word or phrase or action offensive, cultivate your curiosity and look into it. It's trivially easy to type "Why is _____ offensive?" into a search engine and get plentiful responses to help you learn about a topic. Even if you don't end up fully agreeing with the assessment, be humble enough to consider it and curious enough to learn about it. Don't reject unfamiliar ideas out of hand; take some time to consider them and at least deepen your understanding of why your perspective might be different from others' perspectives.

Maintain your curiosity as you engage with media as well. Just as no person is perfect, no piece of media is perfect. Ask questions, and seek out commentaries about the input. How accurate is this biopic? (Some parts were overdramatized, and some parts were actually toned down!) How do LGBTQ+ people feel about this movie's gay romance subplot? (Some hate it, some love it!) Is it just me, or is that character's portrayal kind of racist? (Many people think it is very racist, other people think it is a clever parody of racist caricatures! These different answers beg further questions about what the director's intention for that character was and what its positive or negative impact might be.) **One part of being an agent of change is to become a lifelong learner who asks questions**—preferably with humility and curiosity!

CHAPTER 7 SKILLS SUMMARY

- **Broaden Your Input** to give your brain other ideas, experiences, and information to push back against bias and stereotypes.
- You can do this broadening **via Genuine Connections** with other people, **via Media** you thoughtfully and intentionally consume, and **via Physical, Digital, and Social Representation** in the world around you. Representation matters!
- These are all opportunities to cultivate diverse joy! Although I want everyone to "do the work" to disrupt bias habits, the work *can* be infused with joy!
- Make sure you don't assume that one person's experience speaks for their whole group. Whether it's through genuine connections or media, respect people's experiences as their individual story, without giving it the burden of speaking for others.
- When exposed to new or different ideas, try to **Listen with Humility and Curiosity** as a way to push back against some of the frustrated feelings we know can occur when we're exposed to things that may disrupt our mental habits.

REFLECTION ITEMS

- How diverse is your friends group, across race, gender, sexual orientation, or other group statuses? What are some events or activities you might be interested in that could broaden your input by connecting you with new and different people? Is there an acquaintance or colleague you could text right now to invite for a coffee or to hang out sometime soon?
- What are your favorite types or genres of media? Can you find media in those genres that might broaden your input? With whom might you share that media?

- Reflect on some pieces of media that taught you something about people different from you. What did you learn, and how did it impact you?
- What or who do you often see in your physical, digital, or social environments that might provide opportunities for you to increase representation? What are some causes you believe in that you could display somewhere? What cues do you have, or could you have, in your environment to remind you of your intentions to break bias habits and cultivate diverse joy?
- Can you think of a time you learned about a different way of being or thinking that rubbed you the wrong way at first? How might humility and curiosity help you overcome those feelings in similar situations in the future?
- See if you can generate examples that illustrate how this chapter's content is
  - *actionable* (What are you *doing*?)
  - *self-sustainable* (How will you maintain it over time?)
  - *generalizable* (Consider examples related to race, gender, sexual orientation, age, disability, politics, religion, body size, or other groups.)
  - *customizable* (How might you use these skills in your various work, life, and play contexts?)

SHARE YOUR REFLECTIONS!

Your examples and experiences can contribute to our research, help us improve this training program, and provide guidance to others who complete this training in the future! Consider sharing your examples, insights, and ponderings with us at BiasHabit.com/share. *Related to this chapter, we'd particularly love to hear your recommendations for media that impacted you in some way, so we can share them with others!*

8

TOOLS TO PREVENT BIAS

THE NEXT TWO TOOLS involve things we can do to help prevent bias. To a certain extent, of course, all the tools involve preventing bias. If you use the tools to become more mindful, retrain your mind, and broaden your input, that makes bias less likely, which is another way of saying they prevent bias. After all, the tools are synergistic, and they can all be used in an array of ways and times. The two tools in this chapter—seek individuating information and think ahead—just happen to be a bit more weighted on the side of prevention. They each involve procedures to do ahead of time to prevent bias from even coming into play.

SEEK INDIVIDUATING INFORMATION

When we *don't* know much about someone, generally the main things we do know about them will be things like their apparent race, gender, and age. Without other information about the person in question, those group markers will then exert a greater influence on our perception of that person. Put another way, when we have gaps in our perception of someone, our brains fill those gaps with assumptions taken from stereotypes. When we **Seek Individuating Information**, we learn about and focus on what makes someone unique, and there are fewer gaps for our minds to fill with stereotypes.

This tool is certainly related to broadening your input, as both involve trying to get more information in a way that mitigates the influence of stereotypes and biases. Broadening your input is about *breadth* across various people, media, and representations, while seeking individuating information is about *depth* related to one person.

Imagine that you are a high school teacher and you're trying to learn the names of all twenty of your students. Of the twenty, only one is of Latin American descent, and his name is Juan. Both because he is the *only* representative of his demographic and because his name is related to his ethnicity, it's relatively quick and easy to learn his name. But you also have five blonde girls in your class who, at least superficially, seem very similar. To learn to distinguish them and match the right name with the right girl, you have to try a bit harder. Jenny always wears noticeable earrings, Britney and Daphne both wear glasses, but Britney's are red and Daphne's are black, Courtney asks lots of questions in class, and Tiffany is always late and sits in the back of class. These are still relatively superficial details, but they nevertheless involve you putting in a bit more effort to individuate these girls. In this scenario, all you "need" to distinguish Juan is his ethnicity, but to distinguish the blonde girls, you have to dig into their more unique, individuating details. Think of this scenario as a kind of attentional spotlight as well; because of the demographic patterns in your class, you unintentionally train your attentional processes to pay more attention to the details of the White students.

Let's say now that you need to make some sort of judgment or decision about Juan. All you really know about him is his ethnicity.

Without other, individuating information about Juan in your mind, your brain will be more likely to draw on biases and stereotypes to influence your judgments about him. In contrast, what if you had previously put this tool into practice and gotten to know the details that make Juan a unique individual? Maybe you learn that he is Ecuadorian American; his great-grandparents immigrated to Ohio from Ecuador. He loves soccer—that's a detail that happens to match stereotypes that say Latin American folks are passionate about soccer, but through talking to him, you get to learn it as an actual fact about Juan, rather than an assumption. He's also a violinist, and he likes collecting sneakers. The more your brain encodes deep, unique details you learn about him, the more you see him as an individual. Rather than him being "Juan, the one Latino boy in class," he becomes "Juan, the fourth-generation Ecuadorian American from Ohio who loves soccer, plays the violin, and collects sneakers." Now **it's less likely stereotypes and biases will come into play, because your brain doesn't have as many gaps to fill** in your perception of Juan, and his demographics aren't given as much prominence.

An attendee of my training once pointed out that in this example about Juan, part of what is happening is that maybe we're just bringing in *different* stereotypes—for instance, stereotypes about violinists, soccer fans, and so on. I agree with that assessment! That is *part* of what is happening. Maybe you have stereotypic assumptions about violinists—like that they're sensitive, artistic, delicate, detail oriented, and precise. If *all* you knew about someone was that he was a violinist, you might indeed display some bias habits based on these ideas. But **seeking individuating information is about getting lots of different details, not just one**. Some of these details and stereotypes will contradict each other. Even if we might assume violinists are delicate, we might also assume that soccer fans are aggressive and tough.

In the case of Juan, we now have to resolve those contradictory details—maybe they cancel each other out, leaving us with no expectations about how tough or delicate he is. Also, in the course of getting to know Juan, maybe we see direct evidence of how tough or not tough he is, so the "toughness" stereotypes related to soccer fans or violinists don't even come into play because we don't have a *gap* in our perception that our brains need to try to fill. Once we dig deep into all the details—all the individuating information—about someone, it just makes it less likely that stereotypes and biases even come into play, because **we're seeing that person as an individual**, not "just" as a member of their easily identifiable demographic groups.

Consider this tool in relation to self-fulfilling prophecy. If we lack individuating information, our expectations about a person will likely come from stereotypes related to their group, and our behavior toward them may bring out the behavior we expect. Recall my earlier story about Yoichi, whose new coworkers assumed he would be emotionally cold because he was Japanese and therefore acted coldly toward him, creating a self-fulfilling prophecy. This pattern all changed when another new employee joined the team and, rather than treating Yoichi coldly, really got to know him as an individual. This new employee sought out individuating information, getting to know about Yoichi's family and interests, and learning that, in fact, Yoichi has a very infectious and goofy sense of humor. As others in the work group saw Yoichi and this new friend joking around, they began to see how their assumptions about Yoichi were wrong and they became more friendly! When we have more detailed, specific, individuating information about someone, our expectations will be more attuned to that information rather than stereotypes related to their group memberships.

When you need to make a decision related to someone, for example, in a hiring context, the type of individuating information you seek out might be driven by features of that decision, especially if you have some ideas of specific ways bias might play out in that decision. For instance, in hiring situations, many people report that when they see an applicant with a name that implies an Asian or Latin American ethnicity, assumptions come to mind about that applicant's English proficiency. Maybe they think, "Well, this position requires a lot of writing and speaking, and we need someone with really excellent verbal skills" and then, tacitly assuming that English isn't the applicant's first language, place that applicant lower in their rankings.

The bias in this situation could be prevented by seeking individuating information ahead of time. If spoken English proficiency is an important part of a role you're hiring for, have all applicants submit a speaking sample with their job application. Doing so is seeking individuating information specific to the decision you need to make. If you *know* how well or poorly *every* applicant speaks English because you've heard their speaking samples, then there is no gap for your brain to fill with assumptions based on the ethnicity of their name (or some other factor).

I want to emphasize that, in this example, we're asking *everyone* to submit a speaking sample. We're not singling out people based on our own potential for bias; we're not saying, "Well, let's have all the *Asian* applicants prove their English skills" and making a subset of people do extra work. You're **getting the individuating information about *everyone*.**

On a similar note, someone once told me about a hardware store they worked at, where they had noticed that the manager never hired women for the warehouse side of the store because he assumed that women wouldn't be as physically strong to lift things. After some discussion about this gender bias, they set up a new system. It turned out that the warehouse job description specified that it required the applicant to be able to lift items up to fifty pounds in weight. The hiring manager, however, had just been making assumptions that men could lift that much and women could not. The new system they set up integrated that requirement into the application process; when anyone turned in a job application, the applicant had to show that they could lift a fifty-pound box, and their success or failure was marked on the application. The hiring manager, who at first was a bit unhappy that he was being accused of bias, agreed to this system, and in the end was very happy with the outcome, which involved many more women being hired for the warehouse—and some men being weeded out for those positions. "Ability to lift fifty pounds" is one type of individuating information that was important in this particular role.

As with many of these tools, seeking individuating information can help improve how we interact with *anyone*, not just people from historically disadvantaged groups. In fact, an attendee at one of my trainings, Payal, shared that a type of bias she had been particularly struggling with was bias against White men. Often in realms of diversity discussions, people focus exclusively on bias against people who are members of historically disadvantaged groups—because, of course, our goal is to try to work against that historical pattern of disadvantage. But in terms of the ways our mental biases and assumptions can get in the way of our positive interactions with people, bias can operate in any direction.

In Payal's case, she was an immigrant from India who came to the United States as a scientist. When she got here, many of her colleagues warned her about bias, essentially telling her she needed to "watch out" for White men because they were likely going to be very sexist and very racist toward her. These messages made her feel paralyzed. She believed that surely not *all* White men would treat her poorly, but she didn't know how to take the warnings seriously while also working productively with her many colleagues who were White men. When we talked this issue through, it became clear that seeking individuating information would be a great tool for her. Of course, some people, of any race or gender, might express bias against Payal, but if she could get to know people, seeking individuating information about them, it would make her better able to understand with whom she would have positive interactions and with whom things might be more difficult. Again, this example demonstrates that **the more you get to**

know someone as a unique individual, the less likely it is that you're going to have to rely on assumptions about them.

Benefits in Both Directions

Payal's example also highlights the additional benefits of seeking individuating information, specifically, how it works in both directions. As Payal seeks individuating information about one of her White man colleagues, the process of getting to know him also involves him getting to know *her* better: Through her efforts to get to know him, he also gets more individuating information about her. That individuating information can work against any potential he has to unintentionally express bias toward her.

Think back to the story about the teacher having one Latin student, Juan. As the teacher gets to know all those details of what makes Juan a unique individual, Juan feels seen and feels like his teacher really is invested in him and appreciates him. Maybe he previously had teachers who expressed bias against him, leading him to expect this teacher might as well. But when the teacher really gets to know him, it contradicts his expectation and leads him to engage better with school as well. So again, as with many of these tools, the benefits aren't only for working against bias or preventing bias in our own minds, but also can have additional benefits for the people with whom we're working. If a teacher gets to know individuating information about every single student in their class, maybe those students will perform better and be more engaged with that teacher than if the teacher didn't do those things. **Applying the tools can benefit everyone.**

Putting Individuating Information to Use

Once you've learned individuating information about someone, practice bringing it to the front of your mind when you think about that person. In situations where your mind might quickly categorize someone and move on ("Juan is the one Latin student"), practice drawing on the individuating information to flesh out your mental picture of them as a unique individual ("Juan is the one Latin student and he likes soccer and plays the violin"). In other words, think of this tool both in terms of getting the information in the first place, as a way to get to know someone on a deeper level, and in terms of a mental exercise once you've obtained the information, as a way to put the information to use to practice how you think about them.

I want to encourage you to use this tool proactively, as a way to prevent bias. When you have a new employee, coworker, student, or even just a *potential* new person in your orbit, seeking individuating information is a good way to get to know them and give your mind fewer gaps for stereotypes to fill. In addition to applying this tool proactively as a preventive measure, it can also be used retroactively if you find yourself having stereotypic assumptions come to mind. If you notice your brain is drawing on stereotypes a lot to flesh out your mental image of someone, you can use that as a sign that you don't know them very well. Then you can try to **seek out more individuating information to push back against those stereotypic influences**.

As with all the tools, this tool is synergistic with the others. Seeking individuating information is essentially getting to know someone, as we want to do when we broaden our input via genuine connections. Getting to know what makes them unique will help you succeed as you try to cultivate new friendships. Also, once the friendship forms, the deeper, individuating information will also make the effect of broadening your input stronger, because you have more and more detailed information to push back against the stereotypes and biases. You can also use this tool with broadening your input via the media, as you get to know various characters in media you consume. When you get to know someone more, you'll also be better equipped to consider situational explanations for their behavior and to do perspective-taking, because you'll have a somewhat deeper understanding of their life and circumstances and how they see the world. Individuating information is also something you can reflect on when you detect, reflect, and reject bias and something to rehearse when you rehearse replacements. As a set, these tools work together in our effort to disrupt bias habits and cultivate diverse joy.

THINK AHEAD

Seeking individuating information can also be used in conjunction with our next tool, which is to **Think Ahead**. Bias is more likely when we find ourselves acting spontaneously, when we're surprised or caught in the moment. Those circumstances lead us to rely more on our gut instincts and our brain's automatic, habitual processes. But if we think ahead about how we're going to make a decision or handle a situation, we'll be less vulnerable to showing bias.

Recall the example I used to describe impression justification in chapter 3, in which we imagined selecting a police chief: Our initial gut feelings about gender led us to value credentials more when a man had them than when a woman did. Well, now let's imagine that we think ahead before we look at the the candidates. We can think about how important each kind of credential is for a police chief and then make a commitment to stick to those credentials when judging the applicants. If, ahead of time, we say that we're looking for someone with at least four years of experience, then our "gut feelings" shouldn't color our judgment when we see a candidate who has five years of experience. **When we decide ahead of time how important various criteria are, our decision ends up being based on those criteria**, rather than a story built around our "gut feelings."

This form of thinking ahead is sometimes called *committing to credentials*. You think ahead about what credentials are important for the decision you need to make, then you commit and stick to them when you are in the process of making the decision. Sometimes people hear this advice and interpret it as meaning you should have very strict cutoffs. Some research labs in my field, for instance, have strict grade point average (GPA) cutoffs for students applying to work as research assistants (RAs), and they will not consider anyone below a certain GPA. From a certain point of view, this approach is "fair"—everyone has to meet the same criterion for consideration. But thinking ahead does not necessarily mean your criterion has to be so strict. There can be room for flexible criteria, so long as you've thought it through ahead of time.

Personally, I tend to favor a more holistic approach. I agree that students with high GPAs are likely smart and hardworking. But in my RA hiring process, I never set a minimum GPA. I looked at GPA in conjunction with other factors. The students with very high GPAs were also, on average, more likely to be from affluent families, which meant they were less likely to be working their way through college and in many cases had never held an actual job before. Many of the skills I needed in my RAs happened to overlap with skills they develop when working retail or in service industries. So, if someone had a good-but-not-perfect GPA and also had real job experience, I would invite them for an interview. (Also, we consider lab experience to be a very valuable learning opportunity, and I wanted to make sure potentially less advantaged students had access to that opportunity.)

One other criterion I set was that if someone had a fairly low GPA but they *explained* their low GPA in their application, I always gave that person an interview. To me, that criterion made sense because talking about their low GPA showed that they were aware that their GPA might not compare well to those of others, but they had enough thoughtfulness to try to address it. This criterion came about because of Ronny, one of the very best RAs I ever had. When he applied to work with me, he had a very low GPA, but in the application he specifically explained it, sharing that he had been in a different major that he wasn't suited for. When he applied to my lab, he had just recently changed his major to psychology, where his grades were considerably better. Ronny ended up being one of the best RAs I ever had and even went on to get a PhD in my field!

After hiring Ronny and being so pleased with his amazing contributions to my lab, I thought ahead about my hiring process and made my rule that I would always interview applicants who explained their low GPAs in their application. In the many years since I set that policy, it has led to me hiring many more RAs who were wonderful but couldn't get into other labs that had exclusive GPA requirements. Those labs' losses were my undeniable gain!

The example with Ronny, and how I thought ahead about RA hiring decisions, is a nice example of the difference between equality versus equity. *Equality* involves treating everyone exactly the same—for instance, everyone has to meet the precise GPA cutoff to be considered for a position. My criterion, in which I still consider GPA heavily but also allow other information that might explain a lower GPA, involves more of an equity lens. *Equity* brings into consideration how other factors (e.g., bias, life circumstances) might create disadvantage. My hiring process allowed people to explain how life circumstances might have affected the numerical GPA metric (see also considering situational explanations for behavior). This more equity-conscious criterion still treated everyone equally—everyone was judged based on both GPA *and* the other information. But by thinking ahead and more holistically about how other factors might play into GPA, and by setting my criteria in a way that allowed people to explain their different circumstances and how they contributed to the numerical metric, I was able to work equity into my decision-making process. This example also highlights how you can allow thoughtful flexibility into so-called "objective" metrics, and it invites consideration of whether allegedly "objective" metrics might create inequity, such as with students who have to work full-time jobs to pay for college, which could result in lower GPAs than those of students with more parental financial assistance.

Thinking ahead can also be a way to make sure that policies are applied fairly. For example, a man I'll call Doug experienced a problem with his supervisor showing bias related to requests for time off from work. Doug no longer had a relationship with his conservative Christian family because they didn't approve of him being gay. Whenever his coworkers would request time off for family reasons, the supervisor would readily approve the request because she saw family obligations as important. When Doug put in requests for time off, however, they would receive extra scrutiny ("Do you *really* need to miss work for *that*?"), even when they involved Doug's close friends, whom he considered "chosen family." This bias involves favoring traditional family relationships over chosen family, or even just bias based on someone's life circumstances, lifestyle, or personal choices. We can imagine similar types of bias related to people needing time off for their kids versus people without kids needing time off for another reason.

In this workplace, the supervisor's bias in favor of family events was so apparent and well known that Doug's coworkers started lying to the supervisor, claiming their time off was for family obligations ("I have to travel out of state for my mother's birthday"), even when it was just a vacation with friends. Doug couldn't even get away with lies like that, because the supervisor knew Doug was estranged from his family! Later, when Doug became a supervisor, he thought ahead about this type of bias. He thought that, if anything, he might be a little biased *against* people asking for time off for family reasons because of how his supervisor had treated him. But he didn't *want* to have that sort of bias, so he decided it would be his policy that his employees shouldn't even tell him why they wanted time off. He wanted to remove any possibility of a bias based on his personal subjectivity. He used the Think Ahead tool to consider where and how he might be inclined to show bias, and then set his policy in such a way as to prevent that bias altogether.

I've been using hiring and employment examples for this tool a lot, but it can be applied in other ways as well. I gave our training for a university's housing staff, and during the training session, they identified a way race bias had been playing out, and, through thinking ahead, set a new policy that prevented it. In the dormitory buildings at this university, when residents had deliveries, they could either buzz the delivery person into the building's lobby or they could buzz them all the way into the dormitories to bring the delivery directly to students' rooms. The dormitory front desk staff in the lobby had noticed a pattern, in which students were more likely to buzz White delivery drivers up to their rooms, but Black or brown delivery drivers were more likely to be buzzed only into the lobby. In our discussion, the staff connected this to stereotypes of criminality, as we've discussed several times, and surmised that the residents might be having a bias such that stereotyped expectations made them feel less safe letting Black or brown delivery drivers into the dormitories.

As we thought ahead about this issue, part of what the staff decided was that, really, it was unsafe to let *anyone* unknown all the way into the dormitories. In this case, the bias operated such that residents perhaps felt unjustifiably more safe with White delivery drivers, but really, proper safety dictated no delivery people should be let into the dormitories. This example is another case that shows us that **"bias" doesn't always mean that the way members of historically disadvantaged groups are treated is the "wrong" way**. Careful thought in this case revealed that bias was giving White people extra (and inappropriate) privileges. In other words, by thinking ahead about what is or isn't a safe practice, these housing staff members determined that the correct and safe way to do things was to be equally restrictive with everyone.

Thinking ahead will help ensure you don't fall prey to spontaneous biases in the moment. Habits have a greater influence when we're being more spontaneous. If you can think through what your intentions are and what an ideal outcome would be for a situation you're in, it's more likely your behavior will be aligned with your intentions. I often encourage people to think of the last time they had to have a

> **CONNECTIONS: THINK AHEAD ABOUT REHEARSING REPLACEMENTS**
>
> You can use this tool in conjunction with rehearsing replacements. When I was working on disrupting my habit of calling my students "you guys," I would practice saying "you all" before heading to the classroom. I thought ahead about a change I was trying to make, then rehearsed that change right before I was in the situation where it was needed. If you're entering a situation where you know a particular type of bias is likely to come into play, practice rehearsing or reflecting on examples that counteract that form of bias.

difficult conversation with a significant other, or a child, or a coworker. If you think things through ahead of the conversation—how you want to phrase your position, what you want to convey, what you want the other person to understand, or what sort of information you want from them—it probably goes more smoothly than if you jump into a difficult conversation without thinking it through. Thinking ahead doesn't mean you have to figure out every small conversational detail, but if you can have a clear idea in your head of what you want to happen and some ways things could go wrong (e.g., some ways bias might influence things), it's more likely you'll be able to say or do things more in line with your conscious intentions.

Extended Benefits of Thinking Ahead

The Think Ahead tool can also help us in other ways, because being better prepared in advance just gives us more mental space in general. An educator, Chumani, told me about a time when circumstances forced her to think—and prepare—ahead. She was teaching a course that included both native English speakers and Hmong Americans who were still learning to speak English. The translators working with the Hmong students required Chumani to deliver her teaching materials at least two weeks before each lesson, so they had time to prepare their translations. This practice was annoying for Chumani at first (habit disruption is frustrating!), because she usually did not have her preparations completed that far in advance. But as she got accustomed to this process, she noticed that she was feeling less stress than she usually did when teaching a class, because the advance preparation helped her feel more prepared.

Being less stressed also helps prevent bias, because stress makes our brains more likely to rely on our habits. Recall in chapter 3 when I discussed how supervisors might give people with disabilities fewer opportunities to succeed because they want tasks completed quickly or urgently. Just as in Chumani's story, thinking ahead and doing advance preparation can help prevent this issue. Except for people who work somewhere like an emergency room, most people's work tasks do not have life-or-death urgency. If we know that an employee requires more time to complete a task, we can plan ahead to give them that time, so they have the opportunity to perform up to their potential—rather than allowing our impatience (or lack of preparation) to dictate that tasks go to someone without accommodation needs.

CHAPTER 8 SKILLS SUMMARY

In this chapter, we learned about two tools to help prevent bias.

- We can **Seek Individuating Information** to prevent stereotypes from filling in gaps by focusing on the details that make someone a unique individual. This tool might involve getting to know someone on a deeper, more personal level or perhaps obtaining more information on specific qualifications or past experiences before making a decision about them.
- We can also **Think Ahead** so that we decide ahead of time how to handle a situation, make a decision, or talk about an issue. We decide what criteria are important *before* we're in the situation to make a judgment. Being prepared makes us less likely to fall prey to spontaneous biases.

REFLECTION ITEMS

- When you meet someone new, what are some questions you could ask to get some individuating information about them?
- Consider a time you did or said something wrong because you *didn't* think ahead. How might you have done things differently or better if you had thought things through ahead of time?
- See if you can generate examples that illustrate how this chapter's content is
  - *actionable* (What are you *doing*?)
  - *self-sustainable* (How will you maintain it over time?)
  - *generalizable* (Consider examples related to race, gender, sexual orientation, age, disability, politics, religion, body size, or other groups.)
  - *customizable* (How might you use these skills in your various work, life, and play contexts?)

SHARE YOUR REFLECTIONS!

Your examples and experiences can contribute to our research, help us improve this training program, and provide guidance to others who complete this training in the future! Consider sharing your examples, insights, and ponderings with us at BiasHabit.com/share.

9

TOOLS FOR SPEAKING UP

WHEN BIAS HAPPENS, it can help to speak up about it. When you speak up about bias, it reaffirms your commitment to combating bias, tells others where you stand, and, when done well, can help reduce that kind of bias in others' behavior. The two tools in this chapter will help you with guidelines for how to respond to bias expressed by others and to bias that you've expressed.

SPEAK UP WHEN OTHERS EXPRESS BIAS

I'd like to start our discussion of speaking up by acknowledging that speaking up to others can be difficult! It can be especially difficult to speak up when you're the target of the bias. You might be the one harmed by the bias because someone said something offensive to or about you specifically or toward or about people in your social group generally. If you or your group was the target of bias, your first priority should be your own mental and emotional well-being, and if it happened in a work context, your professional well-being as well. I'm going to share lots of advice about how to speak up effectively, and I'm encouraging *everyone* to speak up about bias, but in doing so I am *not* saying that when you are the one harmed by bias you have to be the bigger person and take on the responsibility of helping the other person learn and grow. **Speaking up should never be an *added* burden on someone who is harmed by bias.** Take care of yourself first!

If you weren't the direct target of bias, or if you were but feel secure enough to speak up, **how you speak up matters** if your goal is to try to create meaningful change. People will often get defensive and not hear what you say if they feel attacked when you speak up. From this book at least, you've learned a lot about different bias habits, but not everyone is aware of how pervasive unintentional bias can be. The research literature gives us a number of guidelines for how to speak up in a way that others are more likely to hear and take to heart. As with all the skills in this book, these guidelines are meant to *enhance*, not replace, your own good sense and your preexisting experience and knowledge about how to have productive conversations with others.

Three Guidelines for Speaking Up

First, speaking up tends to be more effective if **your target is the *behavior*, rather than the *person*.** Targeting the person might sound something like, "You are so racist!" Targeting the behavior might sound like, "When you said *X*, that might have reflected some unintentional bias." Statements targeting the person imply or state that something is fundamentally wrong or immoral inside them, that they are a bad person. When your statement targets the behavior, it allows the other person to separate the undesirable behavior from their sense of self and their view of themself as a fundamentally good person.

I've had some people react to this advice to "let them separate the behavior from their sense of self" by saying, "But that's letting them off the hook!" Yes and no. If you're speaking up at all, you're holding them accountable for their behavior. When you target a specific behavior, that is something they can examine and address (perhaps with your help), but if you target them as a person, they're much more likely to get defensive and disregard what you say. If you truly believe they are fundamentally bigoted, racist, or sexist as a person, targeting them as a person might make *you* feel better, but it likely won't be productive for changing their behavior.

Second, and in a similar vein, speaking up is more effective **if the tone is focused on *working together* rather than *pointing fingers*.** In modern discourse, this concept is sometimes called *calling people in* rather than *calling people out*. A fundamental message of this entire book is that *everyone* is vulnerable to biases, and they're something we *all* can work on. If the other person has done or said something biased, try to imagine whether it's something that you could have ever done: Can you speak up about it by making a statement that includes yourself? When they hear that you might have made a similar statement or done a similar thing, it helps them feel less alone. Perhaps someone feels frustration toward a woman in a leadership role and calls her "bossy" under their breath. You might say something like, "I think societal gender norms make us all vulnerable to devaluing leadership qualities in women. When I have a knee-jerk reaction thinking a woman is 'bossy,' I try to reframe it in a positive light—I like to think she's being assertive, not 'bossy'!" Framing the bias habit as something you have in common and we are all working on together helps the other person not feel singled out. (In fact, I've been applying this advice myself in how I wrote this book, framing bias habits as something we all can work on!) Hearing from someone else that they've made a mistake is hard, and helping them feel less alone in their mistake will make it more likely they hear you and take what you say to heart.

TABLE 3
Guidelines for Speaking Up When Others Express Bias

Do . . .	Do not . . .
Focus on behavior	Focus on the person
Take a collaborative tone	Take an accusatory tone
Focus on concrete, specific instances	Make abstract, general accusations

Third, when speaking up, the content should **focus on *concrete* instances, not *abstract* accusations**. This concept is similar to targeting the behavior rather than the person, but stretches a bit beyond that advice to things like patterns of behavior over time or speaking up about organizational problems. An abstract accusation about a pattern of behavior might sound like, "You're so sexist all the time!," whereas a focus on concrete instances would get specific: "When you said *this*, when you did *this*." With regard to organizations, an abstract accusation might sound like, "This is a fundamentally racist institution!" Even if that is a true statement, it doesn't provide enough concrete information to promote meaningful change. Being more concrete might sound like, "*This* policy creates racial disparities because of *X*, *Y*, and *Z*."

I want to emphasize that these three initial guidelines are *not* making any claims about the objective truth or moral justice of the recommended path. It may be true (1) that the *person* is deeply sexist, (2) that their behavior is uniquely bad and is not something we "all" could work on, or (3) that the institution or pattern of behavior is fundamentally racist. These guidelines address **how best to effect change by speaking up**. Even if deep down, someone is truly, willfully bigoted in their beliefs, your speaking up is more likely to change their behavior if you isolate the behavior, have a tone of working together, and focus on the concrete. Even if the fundamentally bigoted person doesn't change their deep beliefs, your speaking up in these ways can teach them that certain *behaviors* are unacceptable at work, or around you, and lead them to change their behavior accordingly.

I also want to be clear that I'm not telling you how to feel or what you have to believe. It often feels better to call someone out and label them as *racist, sexist, ageist,* or whatever form of bigotry, especially when that's what you believe about them. There are other ways to feel that catharsis; have a chat with a significant other or close friend where you can let loose and label that other person however you wish and get it off your chest. Then, you can use these guidelines to speak up to the perpetrator in a way that is more likely to change their behavior, if that is your goal. I'm also not telling you that you must adopt that goal. **As with all the tools, it's up to you to decide whether to use them and what to do with them.**

Offer Explanations and Viable Solutions

Speaking up is also more effective if you can offer explanations or viable solutions. Throughout this book, you've learned about how and why bias habits operate and how to disrupt those biases. Share that knowledge! Suppose someone has a knee-jerk negative reaction to a new way of doing things or a different type of person than they're used to seeing in a particular role; you can talk to them about norm enforcement and how things "rub us the wrong way" when they differ from what we're used to. Or maybe you're worried about impression justification leading to biases in hiring decisions. You can discuss how it can be helpful to think ahead about what criteria you're looking for, as a way to prevent impression justification.

Just as I've guided us through these concepts on our journey through this book, if they apply to a situation where you're speaking up, you can guide the other person to a deeper understanding of why these biases happen and how to address them. Giving them viable solutions or better ways of doing things in the future is especially useful. If you can offer them a concrete way to improve in the future, they will often latch on to your idea. It gives them a lifeline to say how they can do better next time.

Most People Lack Bad Intentions

When speaking up about bias in others, **acknowledge good intent *and* harmful impact**. One of the best phrases in your journey to speaking up more is "I don't think you *meant* to imply that . . ." You've probably gathered by now that I'm shooing us away from using hot-button words like *racist* and *sexist*—even if they may technically be correct. The term *unintentional bias* is especially useful when speaking up, because *bias* is somewhat less loaded than words like *racist*, and *unintentional* makes it clear that it was likely an accident or mistake. (Related terms, like *implicit*, *automatic*, or *unconscious bias* create other problems, because they can make the bias seem inevitable or hopelessly outside of personal control.)

The vast majority of people don't engage in *intentional* bias. And even for the very small minority of people who are willfully, intentionally biased, the *language* of "unintentional bias" can be useful for behavior change. Even if you suspect, believe, or know that the bias was in fact intentional, acting as though it was "probably unintentional" guides the other person to change their behavior. When you give them the benefit of the doubt by saying that you're sure their *intentions* weren't bad, it leaves them in a position to either correct you, and admit that they had biased intentions, or to go along with you, and acknowledge that they made a mistake. Again, even if it's not what you believe, it's a way to help change their behavior going forward.

Actual *intentional* bias is relatively rare when we look at the proportion of people who have earnest bigoted beliefs. It may not feel that way when we are the targets of intentional bias or when we see its frequency in the news or media, but the vast majority of people you encounter every day believe that all forms of prejudice and bigotry are wrong. For most people you interact with, it is *much* more likely that a biased statement or action was unintentional and arises from them just not realizing how what they said or did may have implications beyond their intentions.

A lot of the time, it's also more mentally healthy for you to just assume that bias was unintentional. When something happens that might involve bias toward you, a mental mantra I encourage you to use is "That had nothing to do with me!" If it was truly an unintentional bias, then it actually *didn't* have much to do with you—it had to do with the bias habits that have been rehearsed in the other person's mind by culture. If the bias was more intentional, it's still healthier for you to distance yourself from it because it reflects the other person's bad intentions, which have no bearing on you or your worth. You can separate yourself from the other person's words or actions and *then* still speak up about them. But do what you can to disallow other people's bad or misguided behavior from disturbing your well-being.

Select the Right Setting

Ideally, address the bias one-on-one with the other person, rather than publicly. People get more defensive when they feel confronted in public. The major exception to this advice is when there is public harm that needs to be addressed. Suppose that in a large meeting, someone says something that implies that women aren't good leaders. You probably want to address that biased statement in the moment, to make it clear to everyone in the room that such statements are not okay. Then you might have a more in-depth conversation one-on-one with the person who made the statement.

Select the Right Messenger

Speaking up will be more effective when it comes from someone with a close relationship to the person who expressed bias. Let's say your coworker Anne says something biased in a meeting. You're not close with Anne, but you're on good terms with her close friend James. Rather than you going to Anne, maybe see if James is comfortable addressing the bias with Anne. Because they have a closer relationship, it's more likely Anne will take what James says to heart. It's always better to address difficult topics with someone you have an established rapport with.

Speaking up will have a bigger impact when it comes from someone who is in a position of authority, like a team leader or supervisor. After all, people in leadership positions are supposed to set the tone for organizational climate and values. In the previous example about someone making a comment implying women aren't good leaders, the person in charge at the meeting might respond to the comment by saying, "Our organization's position is that gender doesn't dictate leadership ability," which uses the leader's authority to state what the organizational standards are. The leader could then *also* address the statement more directly one-on-one with the person who made it (or delegate that task to a supervisor with a close relationship to the person who made the statement).

Also, you have extra power to speak up, and for people to take what you say seriously, if you're not—or not perceived as—a member of the group targeted by the bias. For instance, if a White person speaks up about bias toward an Asian person, or a man speaks up about bias against women, people tend to take it more seriously. There can be a few reasons for this imbalance. If, as a gay man, I speak up about bias against LGBTQ+ people, some people might disregard what I say or give my comments less credence because they see me as "complaining" or being overly sensitive or having selfish motivations for speaking up. But if a straight person brings up the same issue, it's harder to dismiss for those reasons, because the straight person presumably won't "get anything out of it" if the bias is corrected.

An attendee at one of my trainings shared how they experienced this imbalance firsthand. They had been on a workplace diversity committee, during which time they helped write and pass a new policy that included protections for transgender employees. A while after the policy passed, this employee came out as transgender. Their workplace then rescinded the policy, saying that it had been passed under false pretenses because that employee was not "out" as transgender. When they didn't *know* this employee was transgender, they thought the policy would be great, but they changed their tune once they learned the employee was transgender. The merits of a policy should not depend on the identities (known or unknown) of the person proposing it.

Whenever I discuss this imbalance with regard to who is speaking up, I worry that people who *are* members of the targeted group will get discouraged and refrain from speaking up. Let me reiterate that **I want to encourage *everyone* to start speaking up more**. We all need to work together to build a more fair, less biased world. Even if, sometimes, some people might take speaking up less seriously when it comes from a member of the targeted group, the issues you bring to light are still important. And others who hear you speaking up *will* take it to heart and maybe even be more likely to speak up about the same issue themselves. What I am highlighting here is that **allies have *extra* power to be taken seriously** and for their words to effect change. So, everyone, please speak up!

Sometimes allies don't want to speak up because they feel like it's "not their place" to address issues related to an identity they don't have. One idea I want to instill in you is that addressing bias and disparities and cultivating fairness are and should be issues we *all* see as our responsibility. If you don't speak up because you aren't the target of bias, that just means those who are harmed by bias have to do even more work. When you're speaking up as an ally, speak up for values, not "on behalf of" identities that aren't your own. If you're a White person, don't say, "Well, I think Black people would find this offensive." Instead say, "I think this might contribute to racial disparities" or "This falls short of our organization's commitment to racial equity" or "I think this reflects an unintentional racial bias we can address." You're speaking up for your personal or organizational values—for fairness, equity, and justice—not because you think you are a proxy for identities other than your own.

Make sure that your speaking up isn't coming *at the expense of* hearing from members of the targeted group. Don't talk over people or leave it so that targeted group members have no space to speak up, but do use your power as an ally to speak up to make change. I admit that this can be a tricky balance. First off, let me say that, on average, most people who are members of groups targeted by bias are *tired* of speaking up. More often than not, they'll appreciate someone else addressing the issue, rather than feeling that the responsibility is falling on them yet again. But then it can still be awkward: After bias happens, if you want to leave space for members of the targeted group to speak up, but also don't want to leave things unaddressed for so long that they're forced to speak up, how long do you wait before speaking up yourself?

I have a simple solution for this conundrum: **Raise your hand.** That gesture signals that you have something to say but doesn't force your way in to be the first or dominant voice. If a member of the harmed group is exhausted from always speaking up and wishes other people would, they'll be relieved to see your hand up. If a member of the harmed group wants to speak up themselves, they can go ahead and just speak, or they may even follow your example and raise their hand as well. If they've raised their hand, you can then let them speak first.

By definition, bias and inequity will often take the form of having no members of the targeted group present, for instance, in a corporate boardroom where there are no women or no people of color. If we all leave speaking up to members of the disadvantaged group, who will be speaking up in those spaces? Members of those groups are literally not even there to speak! **If fairness and equity are things we all value, then they are things we *all* must speak up for.** (Also, do what you can to bring more underrepresented folks to the table—broaden your input via social representation.)

If there aren't members of the stigmatized group present now, think about how you're speaking up now for members of stigmatized groups who may join your organization later. Also, many stigmatized identities are not necessarily visible—there may be members of targeted groups present who are not, or not yet, comfortable revealing their identities. In those cases, your act of speaking up as an ally can save them from feeling pressure to reveal an identity they may want to keep private.

Using the Tools Together

Speaking up can make use of all the other tools, not only in the content of your speech, such as telling someone about a tool that could help them change the bias, but in how you, as the speaker, approach what you say. When someone does something biased, you can detect it, then reflect on why they did it or where they were coming from, then mentally reject their idea for yourself, and *then* speak up about it. As part of that "reflect" step, maybe you consider situational explanations for their behavior or do perspective-taking to help you understand where the other person is coming from and to figure out when and how to speak up. If you think ahead and figure out some good phrasing for how to speak up about bias, you can rehearse that phrasing from time to time to make it more likely that you automatically say things the way you want to. If you've broadened your input, maybe you can use your experiences with a friend or a media example to describe how or why a type of bias creates harm. If you have more individuating information about the person (e.g., how receptive they are to feedback in general), that can help inform how you speak up.

The thinking ahead tool is especially useful here. Think ahead about what you'd say if you were speaking up about bias, and then you'll be better equipped when you find yourself in a situation that involves speaking up. Is there a particular type of bias you notice frequently in your workplace or life? Think through how you'd like to address it, why it occurs so frequently, and what you'll say the next time you notice it.

SPEAK UP WHEN YOU SLIP UP

It's also important to speak up about your own bias slipups, not just other people's. No one likes the person who's always pointing out everyone else's mistakes but doesn't hold themself accountable. And a core idea throughout this whole process of breaking bias habits is that **everyone will slip up sometimes**. If you do or say something biased that you didn't intend, speak up about it, acknowledge that you displayed bias, and reaffirm your values that oppose bias. Apologize, if that's what's warranted.

When speaking up to others about *their* slipups, I recommended acknowledging good intent first, then harmful impact. When speaking up about your own mistakes, reverse it! If something you said or did was harmful, or you think it might have been harmful, **acknowledge that harm first**, most especially if you're talking to someone who was, or was potentially, harmed by the bias. *After* acknowledging harm, then you can clarify your intentions about what it was you wish you had said or done instead, or how you might make a change for the future. Use your good judgment in the moment (and also, think ahead about how you'd like to address mistakes when they come).

Keep it proportional. Recall the story in chapter 6, when the pastor went way over the top apologizing for misgendering my friend Sandy during their mother's eulogy. That was disproportionate. The big scene the pastor made caused more harm than the initial mistake. Generally speaking, if someone is harmed by something you said or did, they don't want more public attention put on *them* because of *your* mistake. In cases where you believe a big apology is appropriate, doing it one-on-one might be the way to go.

Sometimes people fail to speak up about their mistakes because they're afraid of drawing more attention to their error. They say something a bit biased, then think, "Yikes! I hope no one noticed that" and never address it aloud. If you ever find yourself in that position, I want to assure you that people probably *did* notice. And if you don't address it, all they're left with is what you *did* say or what you *did* do. If you don't speak up about it, they don't know it was a mistake, they don't know it was unintentional, and they don't know you wish you hadn't said it. They're left to think you really believe what you said, with whatever bias or baggage it might imply. You are *much* better served by acknowledging it and correcting yourself.

Also, **when you speak up about your own mistakes, it sets an example that others can follow**. When someone else sees that you addressed your mistake, it gives them a model for how to speak up, which they can rely on the next time they make a mistake. So, as you start speaking up more for yourself, it also starts making it more likely that other people will speak up about their slipups.

CHAPTER 9 SKILLS SUMMARY

Speaking up helps us create change in the world around us.

- If you **Speak Up When Others Express Bias**, do what you can to target the behavior, not the person; set a tone of working together, not finger-pointing; and focus on concrete instances, not abstract accusations.
- When possible, offer explanations or viable solutions.
- Allies and authority figures hold lots of sway when speaking up.
- No one is perfect, so you should also try to **Speak Up When You Slip Up**. Whether speaking up about your own mistakes, those of others around you, or organizational problems, your effort to speak up can positively affect the people around you, who can learn from what you say, as well as follow your example of how to speak up.
- Last and far from least, each time you speak up, whether it's about your own slipup or someone else's, it reinforces the idea that you are the sort of person who speaks up about these things. It reiterates what your values are and how disrupting bias habits is important to you and part of your identity. Each time you speak up, it will get easier to speak up the next time. Practice makes progress!

UNIT 3 WRAP-UP: USE THE TOOLS WISELY

I've emphasized throughout this unit some ways that practicing using one tool can help you get better at using the other tools—**the tools are synergistic**! As with any skill, practice makes progress—the more you employ each tool, the easier and more natural it will feel. With time, the tools can become your new mental habits!

TABLE 4

Tools to Disrupt Bias Habits and Cultivate Diverse Joy

Tool	Definition
Favor mindfulness over blunt, ineffective tools	
Don't suppress stereotypes	Banishing stereotypes from your mind (e.g., "Just try not to stereotype!") backfires.
Don't ignore group statuses	Trying to ignore group membership when interacting with members of different groups (e.g., "I just don't see race!") backfires.
Don't rely on personal objectivity	Bluntly believing that you can be objective when making decisions (e.g., "I'll just be objective!") backfires.
Tools to retrain your mind	
Detect, reflect, and reject bias	Notice biased thoughts or portrayals and label them as biased, think about their origins and consequences, and unemotionally let them go.
Rehearse replacements	Practice rehearsing a new thought or idea to replace or contradict stereotypic thoughts, assumptions, or portrayals.
Consider situational explanations for behavior	Actively consider things outside the person as possible explanations for behavior.
Do perspective-taking	Imagine what it would feel like to be in another person's situation.
Tools to broaden your input	
Broaden your input via genuine connections	Meet different people; diversify your friendships and acquaintanceships.
Broaden your input via media	Expose yourself to movies, books, podcasts, and other media from people different from you.
Broaden your input via physical, digital, and social representation	Increase representation of underrepresented groups wherever you can.
Listen with humility and curiosity	Be eager to learn from others and to question your own way of seeing things.
Tools to prevent bias	
Seek individuating information	Prevent stereotypes from filling in gaps by learning about and focusing on what makes someone unique. Bring these unique details to mind when thinking of that person.

(continued)

TABLE 4 *(continued)*

Tool	Definition
Think ahead	Decide ahead of time how to handle a situation, make a decision, or talk about an issue. Decide what criteria are important before you're in the situation to make a judgment. Being prepared makes you less likely to fall prey to spontaneous biases.
	Tools for speaking up
Speak up when others express bias	Address bias in others by calling them in, giving them the benefit of the doubt, and using the language of "unintentional" bias.
Speak up when you slip up	Be proactive at addressing your own mistakes.

The tools, skills, and advice throughout this book are meant to ***enhance* your own good sense, not *replace* it**. Presumably, you are a functional, good-hearted human being who has a reasonable amount of experience interacting with other people. You know the difference between being kind and being mean. You know your own mind, behavior, and context far better than I ever will. Use these tools wisely to work *with* your own good sense and general human competence.

REFLECTION ITEMS

- Can you think of a time when you or someone else spoke up about bias? How did it go? What do you think contributed to it going well or going poorly?
- Think of some times when you didn't speak up, but wish you had. If you could do things over, how would you have applied what you learned in this chapter to speak up?
- See if you can generate examples that illustrate how this chapter's content is

- *actionable* (What are you *doing*?)
- *self-sustainable* (How will you maintain it over time?)
- *generalizable* (Consider examples related to race, gender, sexual orientation, age, disability, politics, religion, body size, or other groups.)
- *customizable* (How might you use these skills in your various work, life, and play contexts?)

- To wrap up unit 3, consider some additional ways the various tools we learned can be synergistic. How could working on one help you employ one of the others?

SHARE YOUR REFLECTIONS!

Your examples and experiences can contribute to our research, help us improve this training program, and provide guidance to others who complete this training in the future! Consider sharing your examples, insights, and ponderings with us at BiasHabit.com/share.

UNIT FOUR

BE AN AGENT OF PERSONAL AND SOCIAL CHANGE

IN THIS FINAL UNIT, we are going to step back and synthesize everything we've learned on our journey in this book. We will also broaden our scope to see how **the work we do to make personal change** in our own behavior can also be **a powerful force for social change** in the world around us.

10

IT'S UP TO US TO BUILD A BETTER WORLD

CHANGE STARTS WITH YOU. Sometimes people see how big and pervasive problems like racism, sexism, and other forms of prejudice and bigotry are and expect that they cannot make a difference. They think things will only improve when a luminary comes along, like a modern-day Dr. Martin Luther King Jr. They think a transformative leader will come along with big answers, big charisma, and big influence to fix these big problems. They're holding out for a hero to come along and be the agent of change to transform the world. I am here to say, stop waiting. No one is coming. We have to save ourselves. *We* are the agents of change. The world of tomorrow is built by each and every one of us, piece by piece. Meaningful change more often comes from many small actions rather than one big action. Take big actions when you have the power to do so, but you always have the power to make things a little bit better. ***You* are the agent of change.** Everything you do creates ripples affecting the people around you and the future.

UNDERSTAND AND UNDERMINE THE BROAD REACH OF NARROW THINKING

The skills in this book are geared toward helping you be an effective, autonomous agent of change. The awareness skills help us *understand* ways that bias habits play out, and the tools help us *undermine* the influence of those bias habits. What are *you* going to build with the skills and tools you've learned?

Remember that your input determines your output. If you leave things to their own devices, the passive input you get from culture will more often lean into things like stereotypes and biases. But you can take charge of your input! Monitor what messages you're getting from the media or the world around you, broaden your input, and rehearse replacements to take charge and make choices about what you want reinforced in your mind.

Making changes to your input will then bring about different output in the form of thought patterns and behaviors more in line with your intentions. **Remember that biases are habits, and we can disrupt those habits!** When you slip up, return to the steps we learned. Breaking habits involves **Motivation**; creating change has to be something you decide that you care about. It involves **Awareness**, as you come to understand how the bias habits play out. Then you need **Tools** to correct the bias habits and build new habits more in line with your intentions. And then it requires **Effort** over time—*ongoing* effort, as, bit by bit, we work to build a better world.

Nurture mindfulness. Become mindful and aware of biases in your thoughts, expectations, and emotional reactions. When snap judgments and gut feelings come to the surface, remember that they aren't necessarily a reflection of who you are or what you believe—very often, your automatic reactions are just a byproduct of input you've received in life. As you get better at detecting these reactions (and reflecting on and rejecting them!), it will get easier to identify areas you want to work on, and you can bring in many of the tools to help you in those efforts.

When talking about research ideas with our students, my PhD advisor would always say, "These are the best ideas we've had *so far*." What she meant is that anyone can have a better idea, or even just a question that might lead to a better idea. The same goes for the skills in this book. As you work on applying them, you may find or make new and better ways to create meaningful change. Or you'll find ways that one of the skills doesn't quite work for you. Or you'll find a special way that a skill *does* work for you. When any of those things happens, I hope you'll share your experiences with me and my research team! We constantly update our training program, and just as you've benefited from the stories of people who have come before, your stories and insights can help others in the future! Don't be shy! I can use your examples (anonymized, of course) when I give our training live or in future editions of this book. Also, sharing your experiences is another way to solidify what you've learned in your mind. Share your stories and experiences at BiasHabit.com/share, and check out the experiences of other readers who've shared previously at BiasHabit.com/book.

(PERSONAL) PRACTICE MAKES (SOCIAL) PROGRESS

Throughout this book, I've largely kept our attention focused on personal change—on how we as individuals can change our own minds and behaviors. We kept that focus because as individuals we don't always have *direct* influence over institutional or systemic processes, like laws and policies that create disparities, but each and every one of us does have the power to take control of our own behavior. As we've explored these skills, however, I hope you've also started to notice how the work we put into that personal change starts spreading beyond just our own thoughts and behaviors to become social change, starting to influence the world around us.

Recall the example of the employee at the investment firm who identified how self-fulfilling prophecy was creating disparities in their workplace related to which universities new hires had attended, which led to a new policy getting passed. Change in that organization started with the individual looking around and noticing an issue, which inspired a new institutional policy to correct that issue. **That person's actions as an individual resulted in institutional change.** When we broaden our input through making genuine connections with others, our use of that tool likewise *starts* with changing our own behavior, but *then* becomes social change as well as we become more friendly, make more connections with others, and create a more positive climate. All the tools and skills have components of this extension, in which, **as we start doing the work for our personal change, we are also becoming agents of social change** who influence the world around us. As you get better at recognizing and correcting bias within yourself, your power to influence others increases, because your firsthand experience makes you better able to see, understand, and verbalize how bias habits operate in others as well.

Inequity and bias become integrated into social systems and institutions in myriad forms, many of which are unique to specific organizations. This uniqueness is yet another reason that *you* have to be the agent of change. I'm just a scientist-practitioner writing a book; *you* actually know the ins and outs of organizations you're a part of, the day-to-day culture and procedures and policies. Recall the notion of cultural inertia I mentioned in chapter 1. Just like *cognitive* inertia maintains bias habits over time in your mind (chapter 4), *cultural* inertia maintains biases and inequities in cultural systems and institutions. Once something is put *into* a culture—whether that is a movie that becomes part of the zeitgeist or a policy that becomes part of an organization—its default state is to persist. It has inertia, until someone comes along and takes an active role in disrupting it.

At one company that received the bias habit-breaking training, learning about cultural inertia in organizations led an employee to investigate the method used by their company's automatic mail-sorting machines, which automatically marked mail as high or low priority. The machine had been programmed to follow the manual sorting procedures of the previous mailroom staff, who had done what they were taught from the generation before, and so on. After some digging, the employee found out the system was marking mail from Black neighborhoods as low priority—an appalling, racist holdover from an earlier time that had persisted *for decades* due to cultural inertia in their organization. This "institutional bias habit" only came to light because one employee took it upon himself to be an agent of change to investigate this mystery and then correct the obvious problem.

Another important skill to consider as we're talking about institutions is to **Track the Data**. Look at the data you have in your organization that might be relevant to inclusion or equity. What kinds of clients, employees, customers, or members are coming into your organization? Are they staying? What kind of retention numbers do you have? If certain types of folks are leaving at higher rates, what's driving people out? Are people in the same position getting paid the same salary, regardless of gender, race, or other group statuses? Look at whatever kinds of data you can get your hands on that can let you know how your institution is doing with regard to inclusion, rooting out bias, and so on. By tracking the data, you can know where you're doing well, where you're making improvements, and also where you're falling short and where you need to put more attention.

Systemic or institutional change can feel too big for us to tackle as individuals. But **social systems and institutions are *made up of* individuals**, and each of those individuals can create meaningful change. But it can be helpful to think of bias in institutions as also being a type of "habit." Procedures and policies are institutional habits that can likewise be broken with Motivation to change them, Awareness of how they operate, Tools to effectively address them, and institutional Effort. **You have the power to create meaningful change in organizations you are a part of.**

BE A LIFELONG AGENT OF CHANGE

I want you to see yourself as a lifelong agent of personal and social change. The world gets better when each and every one of us is putting in the work to notice and correct problems within ourselves and in the world around us. There are always new things to learn and work on. Find joy in the journey, and discover new people, ideas, and experiences along the way. If you haven't already done so, consider following our monthly podcast *Diverse Joy* as one way to keep up with the work and engage with ongoing learning built on the foundation you started with this book! Share this book, the podcast, or excerpts of what you've learned with others so that you can work on these issues together and provide each other support and feedback in your efforts.

Remember to **approach biases as habits to be broken**, and especially that **breaking those habits requires effort over time**. You've already made a great start by getting through this book; now go forward and apply what you've learned, tuning in to bias habits within your own mind and behavior, and applying tools to disrupt those bias habits and cultivate diverse joy. You can expect that you'll slip up from time to time. There's no magic switch to make bias disappear; even though I've devoted my life to helping people with this work, I still fall prey to bias habits sometimes. **But learn from your mistakes, and return to the skills you've learned to keep making progress.** The more you use the skills you've learned on this journey, the easier they get over time. **Practice makes progress!**

Each and every one of us has the power to make meaningful change, in our own minds, in our behavior, and in the world around us. There will always be naysayers and people who are resistant to change, and there will always be politicians or pundits (on every side) who are ready to pounce on people's passion for or resistance to change and to politicize it for their own gains. Don't let the haters get you down! Another reason this book has focused on the change *you* can make as an individual is because institutions very often let us down and fall short of their alleged commitments to correcting biases and inequity. When institutions inevitably let us down, it falls to us as people to make change in the world. Cut through the noise and focus on what you can influence and the many positive ways you can build a better world around you. Make the work to create change a joyful process. Find joy in the connections you make with others, the progress you make within yourself, and the ways you are opened to new experiences and new ways of seeing the world. If each and every one of us puts in our best efforts to be agents of change, together we can reshape the world. **Today's habits dictate tomorrow's reality.**

Now, take what you've learned and go make meaningful change, in your life and in the world around you!

REFLECTION ITEMS

- Look back through the awareness skills and the tools throughout the book (for help, see table 1 in chapter 4 and table 4 in chapter 9), and see how each of them helps us not only be agents of personal change, but social change as well. Think of how each bias habit can contribute to organizational or systemic problems and how each tool can be used to address those kinds of problems.
- Think about the work you have to do going forward. Remember, breaking bias habits is an ongoing journey, not something that you can check off a list. Keep working to be an agent of change going forward!
- What kinds of data might be important for you to track to see how your organizations are doing with inclusion and equity?
- How will you keep this work going over time? Are there cues you can put in your environment to remind you to keep up the work?
- See if you can generate examples that illustrate how to work on institutional/systemic change that is
    - *actionable* (What are you *doing* in your organization to make meaningful progress?)
    - *self-sustainable* (How will you work on institutional change over time? Are there others in your organization to help you maintain this work?)
    - *generalizable* (How might you address concerns related to various social groups, including race, gender, sexual orientation, age, disability, politics, religion, body size, or other groups.)
    - *customizable* (What are different areas or contexts in your organization where you might work on equity? Think about how customers/clients/employees first come in to your organization, what happens while they are in or working with it, and what happens if or when they leave.)

SHARE YOUR REFLECTIONS!

Your examples and experiences can contribute to our research, help us improve this training program, and provide guidance to others who complete this training in the future! Now, or any time in the future, consider sharing your examples, insights, and ponderings with us at BiasHabit.com/share.

RESOURCES

Check out BiasHabit.com/book for additional resources to supplement this book.

Check out BiasHabit.com/handout for a printable handout that summarizes the core skills from this book!

Follow @BiasHabit on most social media platforms to connect with our nonprofit, to find out when we put out other free resources, and to connect with other people who are on this journey!

If you haven't already, consider following our *Diverse Joy* podcast, as a way to help continue the work started in this book! We designed the podcast specifically to build on and extend the core skills developed in the bias habit-breaking training, in a joyful and entertaining format. Many organizations use the podcast as a monthly diversity discussion activity—you can consider doing the same, at your organization, or with some friends! Through the podcast, we can also go more in-depth on many additional topics beyond what was covered in this book, in addition to each episode revisiting and reminding you about one of the core bias habit-breaking skills. Check it out wherever you get podcasts or at DiverseJoy.com! You can also follow "Diverse Joy" on most social media platforms to see when we put out new content related to the podcast.

References

INTRODUCING OUR JOURNEY: BIAS HABITS AND DIVERSE JOY

Carnes, M.L., Devine, P.G., Manwell, L.B., Byars-Winston, A., Fine, E., Ford, C.E., . . . Sherridan, J. (2015). Effect of an intervention to break the gender bias habit: A cluster randomized, controlled trial. *Academic Medicine, 90*, 221–230.

Cox, W.T.L. (2021). Empowering people to break the bias habit: Bias habit-breaking training review videos [10 videos]. YouTube. http://bit.ly/CoxBHBT

Cox, W.T.L. (2022). Developing scientifically validated bias and diversity trainings that work: Empowering agents of change to reduce bias, create inclusion, and promote equity. *Management Decision, 61*(4), 1038–1061.

Cox, W.T.L., Abramson, L.Y., Devine, P.G., & Hollon, S.D. (2012). Stereotypes, prejudice, and depression: The integrated perspective. *Perspectives on Psychological Science, 7*(5), 427–449.

Cox, W. T. L., & Devine, P. G. (2019). The prejudice habit-breaking intervention: An empowerment-based confrontation approach. In R. K. Mallett & M. J. Monteith (Eds.), *Confronting prejudice and discrimination: The science of changing minds and behaviors* (pp. 249–274). London, United Kingdom: Academic Press.

Cox, W. T. L., & Nelson, A. (2023–present). *Diverse Joy* podcast & educational video series. https://www.biashabit.com/diverse-joy

Devine, P. G., Forscher, P. S., Austin, A. T., & Cox, W. T. L. (2012). Long-term reduction in implicit race bias: A prejudice habit-breaking intervention. *Journal of Experimental Social Psychology*, *4*, 1268–1278.

Devine, P. G., Forscher, P. S., Cox, W. T. L., Sherridan, J., Kaatz, A., & Carnes, M. L. (2017). A gender habit-breaking intervention led to increased hiring of female faculty in STEMM departments. *Journal of Experimental Social Psychology*, *73*, 211–215.

Dobbin, F., & Kalev, A. (2016). Why diversity programs fail. *Harvard Business Review*, *94*, 52–61.

Gino, F., & Coffman, K. (2021). Unconscious bias training that works. *Harvard Business Review*, *99*(5), 114–123.

Greenwald, A. G., & Pettigrew, T. F. (2014). With malice toward none and charity for some: Ingroup favoritism enables discrimination. *American Psychologist*, *69*(7), 669.

Lipman, J. (2018). How diversity training infuriates men and fails women. *Time*, *191*(4), 17–19.

Newkirk, P. (2019). *Diversity Inc.: The failed promise of a billion-dollar business.* New York, NY: Bold Type Books.

Nordell, J. (2017, May 7). Is this how discrimination ends? *The Atlantic*. https://www.theatlantic.com/science/archive/2017/05/unconscious-bias-training/525405/

Shaw, Y., & Matisse, K. M. (2017, June 15). The culture inside [Audio podcast episode]. In *Invisibilia*. NPR. https://www.npr.org/programs/invisibilia/532950995/the-culture-inside

Wen, T. (2020, August 27). *Is it possible to rid police officers of bias?* BBC News. https://www.bbc.com/future/article/20200827-is-it-possible-to-rid-police-officers-of-bias

CHAPTER 1. BIAS HABITS AND HOW TO BREAK THEM

Bigler, R. S., & Liben, L. S. (2007). Developmental intergroup theory: Explaining and reducing children's social stereotyping and prejudice. *Current Directions in Psychological Science, 16*, 162–166. https://doi.org/10.1111/j.1467-8721.2007.00496.x

Canning, M. (2021). *Bias at her fingertips: An interview with Professor Trish Devine.* Duke Corporate Education. https://www.dukece.com/insights/bias-at-her-fingertips/

Charlesworth, T. E., Caliskan, A., & Banaji, M. R. (2022). Historical representations of social groups across 200 years of word embeddings from Google Books. *Proceedings of the National Academy of Sciences, 119*(28), e2121798119.

Charlesworth, T. E., Sanjeev, N., Hatzenbuehler, M. L., & Banaji, M. R. (2023). Identifying and predicting stereotype change in large language corpora: 72 groups, 115 years (1900–2015), and four text sources. *Journal of Personality and Social Psychology, 125*(5), 969.

Cox, W. T. L., & Devine, P. G. (2015). Stereotypes possess heterogeneous directionality: A theoretical and empirical exploration of stereotype structure and content. *PLoS One, 10*(3). https://doi.org/10.1371/journal.pone.0122292

Dancy, T. E. (2014). (Un) Doing hegemony in education: Disrupting school-to-prison pipelines for Black males. *Equity & Excellence in Education, 47*(4), 476–493.

Devine, P. G. (1989). Stereotypes and prejudice: Their automatic and controlled components. *Journal of Personality and Social Psychology, 56*, 5–18.

Dix, E. L., & Devine, P. G. (2024). "It's not an overreaction": Increasing White people's acceptance of the reality of bias and receptivity to Black people's bias concerns. *Journal of Experimental Social Psychology, 110*, 104545.

Dixon, T. L., & Linz, D. (2000). Race and the misrepresentation of victimization on local television news. *Communication Research, 27*(5), 547–573. https://doi.org/10.1177/009365000027005001

Goff, P. A., Jackson, M. C., Di Leone, B. A. L., Culotta, C. M., & DiTomasso, N. A. (2014). The essence of innocence: Consequences of dehumanizing Black children. *Journal of Personality and Social Psychology, 106*(4), 526.

Hofmann, S. G., Asmundson, G. J., & Beck, A. T. (2013). The science of cognitive therapy. *Behavior Therapy, 44*(2), 199–212.

Hollon, S. D. (2003). Does cognitive therapy have an enduring effect? *Cognitive Therapy and Research, 27*, 71–75. https://doi.org/10.1023/A:1022538713914

Hollon, S. D., & Beck, A. T. (2013). Cognitive and cognitive-behavioral therapies. *Bergin and Garfield's Handbook of Psychotherapy and Behavior Change, 6*, 393–442.

Johnson, M. A. (1993, November 29). Crime: New frontier—Jesse Jackson calls it top civil-rights issue. *Chicago Sun-Times*.

Pauker, K., Ambady, N., & Apfelbaum, E. P. (2010). Race salience and essentialist thinking in racial stereotype development. *Child Development, 81*, 1799–1813. https://doi.org/10.1111/j.1467-8624.2010.01511.x

Roselli, D., Matthews, J., & Talagala, N. (2019, May). Managing bias in AI. In *Companion Proceedings of the 2019 World Wide Web Conference* (pp. 539–544). New York, NY: Association for Computing Machinery. https://dl.acm.org/doi/proceedings/10.1145/3308560

Thomas, K. M., Johnson-Bailey, J., Phelps, R. E., Tran, N. M., & Johnson, L. N. (2013). Women of color at midcareer. In L. Comas-Diaz & B. Greene (Eds.), *Psychological health of women of color: Intersections, challenges, and opportunities* (pp. 275–286). Praeger.

Wood, W. (2017). Habit in personality and social psychology. *Personality and Social Psychology Review, 21*(4), 389–403.

Wood, W., & Neal, D. T. (2016). Healthy through habit: Interventions for initiating and maintaining health behavior change. *Behavioral Science and Policy, 2*(1), 71–83.

CHAPTER 2. BIAS HABIT BUILDING BLOCKS: EXPECTATIONS AND EMOTIONS

Berns, G. S., McClure, S. M., Pagnoni, G., & Montague P. R. (2001). Predictability modulates human brain response to reward. *Journal of Neuroscience, 21*(8), 2793–2798. https://doi.org/10.1523/JNEUROSCI.21-08-02793.2001

Berridge, K. C. (2001). Reinforcement, incentives, and expectations. In D. L. Medin (Ed.), *The psychology of learning and motivation* (Vol. 40, pp. 223–278). Academic Press.

Bertrand, M., & Mullainathan, S. (2004). Are Emily and Greg more employable than Lakisha and Jamal? A field experiment on labor market discrimination. *American Economic Review, 94*(4), 991–1013.

Biernat, M. (2009). Stereotypes and shifting standards. In T.D. Nelson (Ed.), *Handbook of prejudice, stereotyping, and discrimination* (pp. 137–152). New York, NY: Psychology Press.

Biernat, M., & Manis, M. (1994). Shifting standards and stereotype-based judgments. *Journal of Personality and Social Psychology, 66*(1), 5.

Brehm, S.S., & Brehm, J.W. (2013). *Psychological reactance: A theory of freedom and control.* New York, NY: Academic Press.

Bruckheim & Patel. (2022, October 13). *Male vs female driving statistics.* https://www.brucklaw.com/male-vs-female-driving-statistics/

Clark, A. (2013). Whatever next? Predictive brains, situated agents, and the future of cognitive science. *Behavioral and Brain Sciences, 36*, 181–253. https://doi.org/10.1017/S0140525X12000477

Devine, P.G., Monteith, M.J., Zuwerink, J.R., & Elliot, A.J. (1991). Prejudice with and without compunction. *Journal of Personality and Social Psychology, 60*(6), 817.

Ferster, C.B., & Skinner, B.F. (1957). *Schedules of reinforcement.* New York, NY: Appleton-Century-Crofts.

Fiorillo, C.D., Tobler, P.N., & Schultz, W. (2003). Discrete coding of reward probability and uncertainty by dopamine neurons. *Science, 299*(5614), 1898–1902. https://doi.org/10.1126/science.1077349

Fontenele, R., & Abi Zeid Daou, M. (2025). Transgender history, part I: An anthropology of gender-nonconformity across ages and cultures. *Behavioral Sciences & the Law, 43*(3), 323–328.

Frank, M.J., Seeberger, L.C., & O'Reilly, R.C. (2004). By carrot or by stick: Cognitive reinforcement learning in parkinsonism. *Science, 306*(5703), 1940–1943. https://doi.org/10.1126/science.1102941

Garrison, J., Erdeniz, B., & Done, J. (2013). Prediction error in reinforcement learning: A meta-analysis of neuroimaging studies. *Neuroscience & Biobehavioral Reviews, 37*(7), 1297–1310. https://doi.org/10.1016/j.neubiorev.2013.03.023

Gilbert, D.T., & Hixon, J.G. (1991). The trouble of thinking: Activation and application of stereotypic beliefs. *Journal of Personality and Social Psychology, 60*, 509–517. https://doi.org/10.1037/0022-3514.60.4.509

Hyde, J.S. (2005). The gender similarities hypothesis. *American Psychologist, 60*, 581–592. https://doi.org/10.1037/0003-066X.60.6.581

Mendes, W., Blascovich, J., Hunter, S., Lickel, B., & Jost, J. (2007). Threatened by the unexpected: Physiological responses during social interactions with expectancy-violating partners. *Journal of Personality and Social Psychology, 92*(4), 698–716. https://doi.org/10.1037/0022-3514.92.4.698

Montague, P. R., Dayan, P., & Sejnowski, T. J. (1996). A framework for mesencephalic dopamine systems based on predictive Hebbian learning. *Journal of Neuroscience, 16*(5), 1936–1947.

Moss-Racusin, C. A., Dovidio, J. F., Brescoll, V. L., Graham, M. J., & Handelsman, J. (2012). Science faculty's subtle gender biases favor male students. *Proceedings of the National Academy of Sciences, 109*, 16474–16479.

Reggev, N., Chowdhary, A., & Mitchell, J. P. (2021). Confirmation of interpersonal expectations is intrinsically rewarding. *Social Cognitive and Affective Neuroscience, 16*(12), 1276–1287.

Schultz, W. (2016). Dopamine reward prediction-error signalling: A two-component response. *Nature Reviews Neuroscience, 17*(3), 183–195.

CHAPTER 3. BIAS HABITS THAT CREATE DISPARITIES

Eagly, A. H., & Koenig, A. M. (2008). Gender prejudice: On the risks of occupying incongruent roles. In E. Borgida & S. T. Fiske (Eds.), *Beyond common sense: Psychological science in the courtroom* (pp. 63–81). Malden, MA: Blackwell.

Nisbett, R. E., & Wilson, T. D. (1977). The halo effect: Evidence for unconscious alteration of judgments. *Journal of Personality and Social Psychology, 35*(4), 250.

Reber, R., Schwarz, N., & Winkielman, P. (2004). Processing fluency and aesthetic pleasure: Is beauty in the perceiver's processing experience? *Personality and Social Psychology Review, 8*(4), 364–382. https://doi.org/10.1207/s15327957pspr0804_3

Rosenthal, R., & Fode, K. L. (1963). The effect of experimenter bias on the performance of the albino rat. *Behavioral Science, 8*(3), 183–189. https://doi.org/10.1002/bs.3830080302

Rosenthal, R., & Jacobson, L. (1968). *Pygmalion in the classroom: Teacher expectation and pupils' intellectual development.* New York, NY: Holt, Rinehart and Winston.

Rudman, L. A., & Fairchild, K. (2004). Reactions to counterstereotypic behavior: The role of backlash in cultural stereotype maintenance. *Journal of Personality and Social Psychology*, *87*(2), 157–176. https://doi.org/10.1037/0022-3514.87.2.157

Thorndike, E. L. (1920). A constant error in psychological ratings. *Journal of Applied Psychology*, *4*(1), 25–29. https://doi.org/10.1037/h0071663

Uhlmann, E. L., & Cohen, G. L. (2005). Constructed criteria: Redefining merit to justify discrimination. *Psychological Science*, *16*(6), 474–480. https://doi.org/10.1111/j.0956-7976.2005.01559.x

Voice of America. (2011, August 14). *Mother of Hubble always aimed for stars*. https://www.voanews.com/a/mother-of-hubble-always-aimed-for-stars--127751383/163252.html

CHAPTER 4. BIAS HABITS THAT PERPETUATE COGNITIVE INERTIA

Antony, J. W., Ferreira, C. S., Norman, K. A., & Wimber, M. (2017). Retrieval as a fast route to memory consolidation. *Trends in Cognitive Sciences*, *21*(8), 573–576. https://doi.org/10.1016/j.tics.2017.05.001

Cox, W. T. L., Devine, P. G., Bischmann, A. A., & Hyde, J. S. (2016). Inferences about sexual orientation: The roles of stereotypes, faces, and the gaydar myth. *Journal of Sex Research*, *53*(2), 157–171. https://doi.org/10.1080/00224499.2015.1015714

Cox, W. T. L., Devine, P. G., Bischmann, A. A., & Hyde, J. S. (2017). Ecological invalidity of existing gaydar research: In-lab accuracy translates to real-world inaccuracy: Response to Rule, Johnson, & Freeman (2016). *Journal of Sex Research*, *54*(7), 820–824.

Cox, W. T. L., Xie, X., & Devine, P. G. (2022). Untested assumptions perpetuate stereotyping: Learning in the absence of evidence. *Journal of Experimental Social Psychology*, *102*, 1–13. https://doi.org/10.1016/j.jesp.2022.104380.

Darley, J. M., & Gross, P. H. (1983). A hypothesis-confirming bias in labeling effects. *Journal of Personality and Social Psychology*, *44*(1), 20.

Eberhardt, J. L., Goff, P. A., Purdie, V. J., & Davies, P. G. (2004). Seeing black: Race, crime, and visual processing. *Journal of Personality and Social Psychology*, *87*(6), 876.

Gibson, B. R., Rogers, T. T., & Zhu, X. (2013). Human semi-supervised learning. *Topics in Cognitive Science*, *5*(1), 132–172. https://doi.org/10.1111/tops.12010

Gilbert, D. T. (1991). How mental systems believe. *American Psychologist, 46*(2), 107. https://doi.org/10.1037/0003-066X.46.2.107

Gilbert, D. T., Tafarodi, R. W., & Malone, P. S. (1993). You can't not believe everything you read. *Journal of Personality and Social Psychology, 65*(2), 221. https://doi.org/10.1037/0022-3514.65.2.221

Hebb, D. O. (1949). *The organization of behavior: A neuropsychological approach*. New York, NY: Wiley.

Kalish, C. W., Rogers, T. T., Lang, J., & Zhu, X. (2011). Can semi-supervised learning explain incorrect beliefs about categories? *Cognition, 120*(1), 106–118.

Lord, C. G., Ross, L., & Lepper, M. R. (1979). Biased assimilation and attitude polarization: The effects of prior theories on subsequently considered evidence. *Journal of Personality and Social Psychology, 37*(11), 2098. https://doi.org/10.1037/0022-3514.37.11.2098

McClelland, J. L. (2006). How far can you go with Hebbian learning, and when does it lead you astray? In Y. Munakata & M. H. Johnson (Eds.), *Processes of a change in brain and cognition development: Attention and performance* (pp. 36–69). Oxford University Press.

McClelland, J., Botvinick, M., Noelle, D., Plaut, D., Rogers, T., Seidenberg, M., & Smith, L. (2010). Letting structure emerge: Connectionist and dynamical systems approaches to cognition. *Trends in Cognitive Sciences, 14*(8), 348–356.

Nickerson, R. S. (1998). Confirmation bias: A ubiquitous phenomenon in many guises. *Review of General Psychology, 2*(2), 175–220. https://doi.org/10.1037/1089-2680.2.2.175

Pohl, R. (Ed.). (2004). *Cognitive illusions: A handbook on fallacies and biases in thinking, judgement and memory*. Psychology Press.

Reeves, A. N. (2014). *Written in black & white: Exploring confirmation bias in racialized perceptions of writing skills*. Nextions. https://nextions.com/insights/perspectives/written-in-black-white-exploring-confirmation-bias-in-racialized-perceptions-of-writing-skills/

Schwarz, N., Newman, E., & Leach, W. (2016). Making the truth stick & the myths fade: Lessons from cognitive psychology. *Behavioral Science & Policy, 2*(1), 85–95.

CHAPTER 5. FAVOR MINDFULNESS OVER BLUNT, INEFFECTIVE TOOLS

Apfelbaum, E. P., Norton, M. I., & Sommers, S. R. (2012). Racial colorblindness: Emergence, practice, and implications. *Current Directions in Psychological Science, 21*, 205–209.

Apfelbaum, E. P., Sommers, S. R., & Norton, M. I. (2008). Seeing race and seeming racist? Evaluating strategic colorblindness in social interaction. *Journal of Personality and Social Psychology, 95*, 918–932.

Creswell, J. D. (2017). Mindfulness interventions. *Annual Review of Psychology, 68*(1), 491–516.

Fuochi, G., Boin, J., Lucarini, A., & Voci, A. (2023). A mindful path toward prejudice reduction: Key mindfulness facets and mediators for promoting positive intergroup relations. *Mindfulness, 14*(12), 2894–2905.

Harris, R. (2006). Embracing your demons: An overview of acceptance and commitment therapy. *Psychotherapy in Australia, 12*(4), 70–76.

Kulick, C., Perry, E., & Bourhis, A. (2000). Ironic evaluation processes: Effects of thought suppression on evaluations of older job applicants. *Journal of Organizational Behaviour, 21*(6), 689–711.

Lueke, A., & Gibson, B. (2016). Brief mindfulness meditation reduces discrimination. *Psychology of Consciousness: Theory, Research, and Practice, 3*(1), 34.

Macrae, C. N., Bodenhausen, G. V., Milne, A. B., & Jetten, J. (1994). Out of mind but back in sight: Stereotypes on the rebound. *Journal of Personality and Social Psychology, 67*(5), 808.

Norton, M. I., Sommers, S. R., Apfelbaum, E. P., Pura, N., & Ariely, D. (2006). Color blindness and interracial interaction: Playing the political correctness game. *Psychological Science, 17*, 949–953.

Plaut, V. C., Thomas, K. M., & Goren, M. J. (2009). Is multiculturalism or color blindness better for minorities? *Psychological Science, 20*(4), 444–446. https://doi.org/10.1111/j.1467-9280.2009.02318.x

Shapiro, S. L., Carlson, L. E., Astin, J. A., & Freedman, B. (2006). Mechanisms of mindfulness. *Journal of Clinical Psychology, 62*(3), 373–386.

Uhlmann, E. L., & Cohen, G. L. (2007). "I think it, therefore it's true": Effects of self-perceived objectivity on hiring discrimination. *Organizational Behavior and Human Decision Processes, 104*, 207–223. https://doi.org/10.1016/j.obhdp.2007.07.001

Wegner, D. M., Schneider, D. J., Carter, S. R., & White, T. L. (1987). Paradoxical effects of thought suppression. *Journal of Personality and Social Psychology, 53*(1), 5–13. https://doi.org/10.1037//0022-3514.53.1.5

CHAPTER 6. TOOLS TO RETRAIN YOUR MIND

Abramson, L. Y., Seligman, M. E., & Teasdale, J. D. (1978). Learned helplessness in humans: Critique and reformulation. *Journal of Abnormal Psychology, 87*(1), 49.

Avery, D. R., McKay, P. F., & Wilson, D. C. (2008). What are the odds? How demographic similarity affects the prevalence of perceived employment discrimination. *Journal of Applied Psychology, 93*(2), 235.

Beck, J. S. (2021). *Cognitive therapy: Basics and beyond* (3rd ed.). New York, NY: Guilford Press.

Devine, P. G., & Monteith, M. J. (1993). The role of discrepancy-associated affect in prejudice reduction. In D. M. Mackie & D. L. Hamilton (Eds.), *Affect, cognition and stereotyping* (pp. 317–344). Academic Press.

Galinsky, A. D., & Moskowitz, G. B. (2000). Perspective-taking: Decreasing stereotype expression, stereotype accessibility, and in-group favoritism. *Journal of Personality and Social Psychology, 78*(4), 708–724. https://doi.org/10.1037/0022-3514.78.4.708

Ku, G., Wang, C. S., & Galinsky, A. D. (2015). The promise and perversity of perspective-taking in organizations. *Research in Organizational Behavior, 35*, 79–102.

Kurdi, B., Sanchez, A., Dasgupta, N., & Banaji, M. R. (2023). (When) do counterattitudinal exemplars shift implicit racial evaluations? Replications and extensions of Dasgupta and Greenwald (2001). *Journal of Personality and Social Psychology: Attitude and Social Cognition, 126*(4), 543–565.

Monteith, M. J. (1993). Self-regulation of prejudiced responses: Implications for progress in prejudice-reduction efforts. *Journal of Personality and Social Psychology, 65*, 469–485. https://doi.org/10.1037/0022-3514.65.3.469

Monteith, M. J., Ashburn-Nardo, L., Voils, C. I., & Czopp, A. M. (2002). Putting the brakes on prejudice: On the development and operation of cues for control. *Journal of Personality and Social Psychology, 83*, 1029–1050.

Monteith, M. J., Voils, C. I., & Ashburn-Nardo, L. (2001). Taking a look underground: Detecting, interpreting, and reacting to implicit racial biases. *Social Cognition, 19*, 395–417.

Stewart, T.L., La Duke, J.R., Bracht, C., Sweet, B.A.M., & Gamarel, K.E. (2003). Do the eyes have it? A program evaluation of Jane Elliott's Blue eyes/Brown eyes diversity training exercise. *Journal of Applied Social Psychology, 33*, 1898–1921.

Stewart, T.L., Latu, I.M., Kawakami, K., & Myers, A.C. (2010). Consider the situation: Reducing automatic stereotyping through situational attribution training. *Journal of Experimental Social Psychology, 46*(1), 221–225.

CHAPTER 7. TOOLS TO BROADEN YOUR INPUT

Allport, G.W. (1954). *The nature of prejudice*. Reading, MA: Addison-Wesley.

Avery, D.R., Volpone, S.D., Stewart, R.W., Luksyte, A., Hernandez, M., McKay, P.F., & Hebl, M.M.R. (2013). Examining the draw of diversity: How diversity climate perceptions affect job-pursuit intentions. *Human Resource Management, 52*(2), 175–193.

Cheryan, S., Plaut, V.C., Davies, P.G., & Steele, C.M. (2009). Ambient belonging: How stereotypical cues impact gender participation in computer science. *Journal of Personality and Social Psychology, 97*(6), 1045–1060. https://doi.org/10.1037/a0016239

Eibach, R.P., & Ehrlinger, J. (2006). "Keep your eyes on the prize": Reference points and racial differences in assessing progress toward equality. *Personality and Social Psychology Bulletin, 32*(1), 66–77.

Er-Rafiy, A., Brauer, M., & Musca, S.C. (2010). Effective reduction of prejudice and discrimination: Methodological considerations and three field experiments. *Revue internationale de psychologie sociale, 23*(2), 57–95.

Golubovich, J., & Ryan, A.M. (2022). Implications of diversity cues in recruitment and assessment materials: Reactions and performance. *International Journal of Selection and Assessment, 30*(4), 467–485.

Kalla, J.L., & Broockman, D.E. (2023). Which narrative strategies durably reduce prejudice? Evidence from field and survey experiments supporting the efficacy of perspective-getting. *American Journal of Political Science, 67*(1), 185–204.

Kawakami, K., Dovidio, J.F., Moll, J., Hermsen, S., & Russin, A. (2000). Just say no (to stereotyping): Effects of training in the negation of stereotypic associations on stereotype activation. *Journal of Personality and Social Psychology, 78*(5), 871.

Kimmelman, P. (2013, November 8). Banaji reveals biases in talk. *Yale Daily News.* https://yaledailynews.com/blog/2013/11/08/banaji-reveals-biases-in-talk/

Murrar, S., & Brauer, M. (2018). Entertainment-education effectively reduces prejudice. *Group Processes & Intergroup Relations, 21*(7), 1053–1077.

Page-Gould, E., Mendoza-Denton, R., & Tropp, L. R. (2008). With a little help from my cross-group friend: Reducing anxiety in intergroup contexts through cross-group friendship. *Journal of Personality and Social Psychology, 95*, 1080–1094. https://doi.org/10.1037/0022-3514.95.5.1080

Pettigrew, T. F., & Tropp, L. R. (2006). A meta-analytic test of intergroup contact theory. *Journal of Personality and Social Psychology, 90*(5), 751–783. https://doi.org/10.1037/0022-3514.90.5.751

Ryan, C. S., Judd, C. M., & Park, B. (1996). Effects of racial stereotypes on judgments of individuals: The moderating role of perceived group variability. *Journal of Experimental Social Psychology, 32*(1), 71–103.

Shelton, J. N., & Richeson, J. A. (2005). Intergroup contact and pluralistic ignorance. *Journal of Personality and Social Psychology, 88*(1), 91–107. https://doi.org/10.1037/0022–3514.88.1.91

Twain, M. (1869). *The innocents abroad.* H. H. Bancroft and Company; American Publishing Company.

CHAPTER 8. TOOLS TO PREVENT BIAS

Brewer, M. B. (1988). A dual process model of impression formation. In T. K. Srull & R. S. Wyer, Jr. (Eds.), *Advances in social cognition* (Vol. 1, pp. 1–36). Hillsdale, NJ: Lawrence Erlbaum Associates.

Fiske, S. T., & Neuberg, S. L. (1990). A continuum of impression formation, from category-based to individuating processes: Influences of information and motivation on attention and interpretation. *Advances in Experimental Social Psychology, 23*, 1–74. https://doi.org/10.1016/s0065-2601(08)60317-2

Hugenberg, K., Miller, J., & Claypool, H. M. (2007). Categorization and individuation in the cross-race recognition deficit: Toward a solution to an insidious problem. *Journal of Experimental Social Psychology, 43*(2), 334–340.

Meissner, P., & Wulf, T. (2013). Cognitive benefits of scenario planning: Its impact on biases and decision quality. *Technological Forecasting and Social Change, 80*(4), 801–814.

Qian, M. K., Quinn, P. C., Heyman, G. D., Pascalis, O., Fu, G., & Lee, K. (2019). A long-term effect of perceptual individuation training on reducing implicit racial bias in preschool children. *Child Development*, *90*(3), e290–e305.

Quinn, D. M. (2020). Experimental evidence on teachers' racial bias in student evaluation: The role of grading scales. *Educational Evaluation and Policy Analysis*, *42*(3), 375–392.

Uhlmann, E. L., & Cohen, G. L. (2005). Constructed criteria: Redefining merit to justify discrimination. *Psychological Science*, *16*(6), 474–480. https://doi.org/10.1111/j.0956-7976.2005.01559.x

CHAPTER 9. TOOLS FOR SPEAKING UP

Cox, W. T. L., & Devine, P. G. (2025). Changing implicit bias vs empowering people to address the personal dilemma of unintentional bias. In J. A. Krosnick, T. H. Stark, & A. L. Scott (Eds.), *The Cambridge handbook of implicit bias and racism* (pp. 563–571). Cambridge, MA: Cambridge University Press.

Czopp, A. M., & Monteith, M. J. (2003). Confronting prejudice (literally): Reactions to confrontations of racial and gender bias. *Personality and Social Psychology Bulletin*, *29*(4), 532–544. https://doi.org/10.1177/0146167202250923

Czopp, A. M., Monteith, M. J., & Mark, A. Y. (2006). Standing up for a change: Reducing bias through interpersonal confrontation. *Journal of Personality and Social Psychology*, *90*(5), 784.

Forscher, P. S. (2016). *The individually-targeted habit-breaking intervention and group-level change* [Doctoral dissertation, University of Wisconsin–Madison]. Thesis Commons. https://osf.io/preprints/thesiscommons/4t7fy_v1

Forscher, P. S., Cox, W. T. L., Graetz, N., & Devine, P. G. (2015). The motivation to express prejudice. *Journal of Personality and Social Psychology*, *109*, 791–812.

Forscher, P. S., Mitamura, C., Cox, W. T. L., Dix, E. L., & Devine, P. G. (2017). Breaking the prejudice habit: Mechanisms, timecourse, and longevity. *Journal of Experimental Social Psychology*, *72*, 133–146.

Gardner, D. M., & Ryan, A. M. (2020). What's in it for you? Demographics and self-interest perceptions in diversity promotion. *Journal of Applied Psychology*, *105*(9), 1062.

Howe, L. C., Carr, P. B., & Walton, G. M. (2022). Normative appeals motivate people to contribute to collective action problems more when they invite people to work together toward a common goal. *Journal of Personality and Social Psychology, 121*(2), 215–238. https://doi.org/10.1037/pspa0000278

Kaiser, C. R., & Miller, C. T. (2001). Stop complaining! The social costs of making attributions to discrimination. *Personality and Social Psychology Bulletin, 27*, 254–263.

Kaiser, C. R., & Miller, C. T. (2004). A stress and coping perspective on confronting sexism. *Psychology of Women Quarterly, 28*, 168–178.

King, E. B., Hebl, M., Shapiro, J. R., Silver, E., Bilotta, I., Lennon, N., . . . Corrington, A. (2024). (Absent) allyship in STEM: Can psychological standing increase prejudice confrontation? *Journal of Business and Psychology, 39*(4), 1–21.

Petty, R. E., Fleming, M. A., Priester, J. R., & Feinstein, A. H. (2001). Individual versus group interest violation: Surprise as a determinant of argument scrutiny and persuasion. *Social Cognition, 19*, 418–442.

Plant, E. A., & Devine, P. G. (2009). The active control of prejudice: Unpacking the intentions guiding control efforts. *Journal of Personality and Social Psychology, 96*, 640–652.

Sabat, I., Trump, R., & King, E. (2014). Individual, interpersonal, and contextual factors relating to disclosure decisions of lesbian, gay, and bisexual individuals. *Psychology of Sexual Orientation and Gender Diversity, 1*(4), 431.

Schultz, J. R., & Maddox, K. B. (2013). Shooting the messenger to spite the message? Exploring reactions to claims of racial bias. *Personality and Social Psychology Bulletin, 39*(3), 346–358.

Woodzicka, J. A., Mallett, R. K., Hendricks, S., & Pruitt, A. V. (2015). It's just a (sexist) joke: Comparing reactions to sexist versus racist communications. *Humor, 28*(2), 289–309.

Founded in 1893,
UNIVERSITY OF CALIFORNIA PRESS
publishes bold, progressive books and journals on topics in the arts, humanities, social sciences, and natural sciences—with a focus on social justice issues—that inspire thought and action among readers worldwide.

The UC PRESS FOUNDATION
raises funds to uphold the press's vital role as an independent, nonprofit publisher, and receives philanthropic support from a wide range of individuals and institutions—and from committed readers like you. To learn more, visit ucpress.edu/supportus.